Procedures

Electrical Installation Series – Foundation Course

Ted Stocks
Malcolm Doughton
Charles Duncan
Thomas Awcock

Edited by Chris Cox

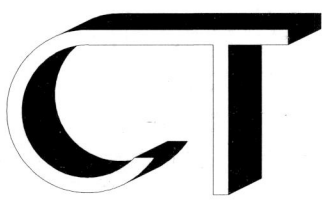

MACMILLAN

First published 1998 by
MACMILLAN PRESS LTD
Houndmills, Basingstoke, Hampshire RG21 6XS
and London
Companies and representatives throughout the world

ISBN 0-333-71986-7

A catalogue record for this book is available from the
British Library.

This book is printed on paper suitable for recycling and
made from fully managed and sustained forest sources.

10   9   8   7   6   5   4   3   2   1
07 06 05 04 03 02 01 00 99 98

Printed in Great Britain by L&S Printing Co. Ltd

The publisher, authors and CT Projects believe that the
information contained in this book is correct at the time of
going to press. All parties making use of this information must
rely on their own skill and judgement and take all precautions
regarding health and safety to avoid any loss or damage. The
authors and CT Projects assume no liability to anyone for any
loss or damage caused by error or omission in this book.

About this book

"Procedures" is one of a series of books published by Macmillan Press Ltd related to Electrical Installation Work. The series may be used to form part of a recognised course, for example City and Guilds Course 2360, or individual books can be used to update knowledge within particular subject areas. A complete list of titles in the series is given below.

Foundation Course books give the student the underpinning knowledge criteria required for City and Guilds Course 2360 Part I Theory. The supplementary book, *Practical Requirements and Exercises*, covers the additional underpinning knowledge required for the Part I Practice.

Level 2 NVQ

Candidates who successfully complete assignments towards the City and Guilds 2360 Theory and/or Practice Part I can apply this success towards Level 2 NVQ through a process of Accreditation of Prior Learning.

Electrical Installation Series

Foundation Course
Starting Work
Procedures
Basic Science and Electronics

Supplementary title:
Practical Requirements and Exercises

Intermediate Course
The Importance of Quality
Stage 1 Design
Intermediate Science and Theory

Supplementary title:
Practical Tasks

Advanced Course
Advanced Science
Stage 2 Design
Electrical Machines
Lighting Systems
Supplying Installations

Acknowledgements

The authors and publishers gratefully acknowledge the following illustration sources:

Maplin Electronics for Figure 2.21

RS Components for Figures 6.36, 7.2

Every effort has been made to trace all copyright holders but if any have been inadvertently overlooked, the publishers will be pleased to make the necessary arrangements at the first opportunity.

Study guide

This studybook has been written to enable you to study either in a classroom or in an open or distance learning situation. To ensure that you gain the maximum benefit from the material you will find prompts all the way through that are designed to keep you involved with the subject. The book has been divided into 24 parts each of which may be suitable as one lesson in the classroom situation. However if you are studying by yourself the following points may help you.

☞ Work out when, and for how long, you can study each week. Complete the table below and from this produce a programme so that you will know approximately when you should complete each chapter and take the progress and end tests. Your tutor may be able to help you with this. It may be necessary to reassess this timetable from time to time according to your situation.

☞ Try not to take on too much studying at a time. Limit yourself to between 1 hour and 2 hours and finish with a task or the self assessment questions (SAQ). When you resume your study go over this same piece of work before you start a new topic.

☞ You will find the answers to the questions at the back of this book, but before you look at the answers check that you have read and understood the question and written the answer you intended.

☞ A "progress check" at the end of Chapter 5 and an "end test" covering all the material in this book are included so that you can assess your progress.

☞ Tasks are included where you are given the opportunity to ask colleagues at work or your tutor at college questions about practical aspects of the subject. There are also tasks where you may be required to use manufacturers' catalogues to look answers up. These are all important and will aid your understanding of the subject.

☞ You will need to have available for reference current copies of IEE Guidance Note 1, BS 7671, and a cable manufacturer's catalogue in order to complete some of the exercises within this book. You will be prompted at the beginning of any chapter where this applies so that you can obtain the relevant material before you start. There are also times when it would be helpful to be able to refer to IEE Guidance Note 3.

☞ Your safety is of paramount importance. You are expected to adhere at all times to current regulations, recommendations and guidelines for health and safety.

Study times					
	a.m. from	to	p.m. from	to	Total
Monday					
Tuesday					
Wednesday					
Thursday					
Friday					
Saturday					
Sunday					

Programme	Date to be achieved by
Chapter 1	
Chapter 2	
Chapter 3	
Chapter 4	
Chapter 5	
Progress check	
Chapter 6	
Chapter 7	
Chapter 8	
Chapter 9	
Chapter 10	
End test	

Contents

1

Electricity Suppliers' Tariffs

You will need to have available for reference current copies of IEE Guidance Note 1, BS 7671, and a cable manufacturer's catalogue in order to complete some of the exercises within this book. You will be prompted at the beginning of any chapter where this applies so that you can obtain the relevant material before you start. There are also times when it could be helpful to be able to refer to a current copy of Guidance Note 3. You will not require them for this chapter.

At the beginning of all the other chapters in this book you will be asked to complete a short revision exercise based on the previous chapter – Sid with a clipboard will remind you of this. Before we begin this chapter remind yourself of the following facts from *Starting Work*.

What voltages are likely to be used by

domestic property?

light industry?

heavy industry?

What is the voltage limit for portable handheld equipment on a constuction site?

How should cable on card drums be stored?

How should accessories be stored?

On completion of this chapter you should be able to:

◆ recognise the difference between normal and restricted tariffs
◆ calculate an electricity usage cost from different factors
◆ explain the relationship between the phases of a three-phase supply
◆ determine the relationship between voltages in star and delta connected supplies
◆ complete the revision exercise at the beginning of the next chapter

In the first book in this series, *Starting Work*, we looked at the generation and distribution of electricity. After the supply has been distributed it will end up at the consumer's main supply intake.

All installations are metered so that payment can be calculated. The energy, or kilowatt hour (kWh) meter, is connected between the supply and all of the consumer's equipment to ensure that the supply company charges for all the electricity used by the consumer (Figure 1.1).

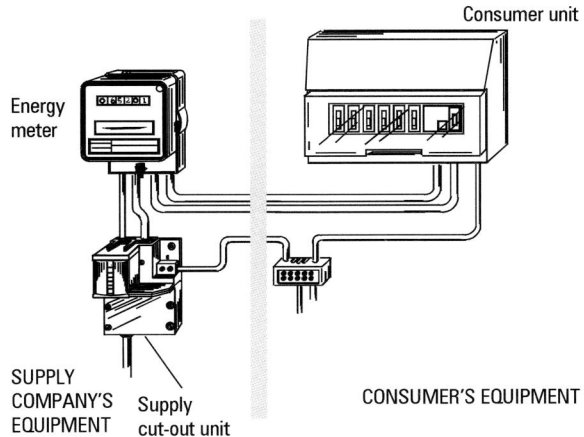

Figure 1.1   *Domestic intake position showing the position of the energy meter*

Energy

All energy is measured using the same unit, be it burning coal or gas, or energy in the human body. The SI unit of energy is the joule. One joule is equal to one watt of power used in one second. When measuring the amount of energy used in a house or factory a joule is far too small a unit, so the power is multiplied up into kilowatts and the time into hours. So the unit used for costing electrical energy is the kilowatt hour (kWh).

1 kWh   = 3 600 000 joules
(60 seconds × 60 minutes × 1000 watts)

A 1 kW electric heater when left on continuously for one hour uses one unit of electricity. Electricity is costed on the basis of units or kWh.

1 unit of electricity = 1 kWh

The "electricity" meter that measures the amount of electrical energy used is calibrated in units. Each unit is given a costing by the area electricity boards and from this a bill is raised.

The cost of electricity

Electricity is not used at the same rate throughout a 24 hour day (Figure 1.2). This is reflected in the price the area electricity boards have to pay the generating companies for the electricity. When there is a peak demand the cost is far greater than when there is a trough. These differences in costs are passed on to the customer if they choose one of the economy types of tariff.

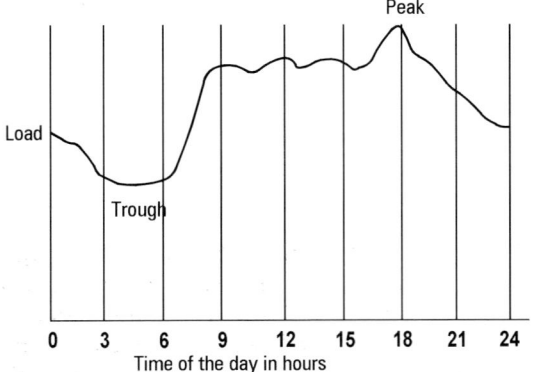

Figure 1.2   The load demand over 24 hours

For domestic consumers there are two main tariffs available:
- the Economy 7 rate
- the Standard rate

The Economy 7 rate

This is designed for consumers that use some electrical equipment at night during the low consumption times. During a seven hour period, usually somewhere between 11 p.m. and 7 a.m., the cost of electricity to the consumer is about half that of any other time.

Domestic consumers can use their washing machines, tumble dryers or dishwashers during this "off-peak" time and take advantage of the cheaper rate. A time switch installed at the appliance ensures that the equipment switches on automatically without someone having to wait up half the night to switch it on (Figure 1.3).

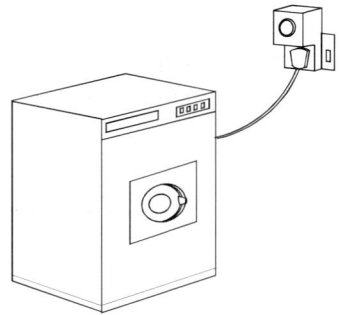

Figure 1.3

So that the consumer is charged at the correct rate for the time of day that the electricity is used, a dual tariff meter is installed by the supply authority.

The Standard rate

This is the traditional rate that uses a single tariff meter that records all of the electrical energy used regardless of the time of day or night. The price per unit of the electricity supplied throughout the twenty-four hour period is the same all the time.

Make-up of tariffs

Most tariffs are made up of two main costs:
- a standing quarterly charge
- a charge for each unit supplied.

Standing quarterly charge

These charges are designed to cover, in part, those elements of costs which do not vary with consumption. The main items include:
- meter reading
- supplying and maintaining service cables, service terminations and meters
- billing and collection of accounts
- consumer service, maintaining records and answering account queries.

In the Economy 7 "off-peak" tariff the quarterly charge is slightly higher, reflecting the costs of the more complex metering and switching.

Unit charge

This is a set price for every unit used. One unit is one kilowatt used for one hour (1 kWh). The cost of one unit depends on the particular tariff in use. On a Standard rate the cost per unit is the same throughout the day and night. However, with the Economy 7 "off-peak" rate, for seven hours at night the price per unit is less than half of that during the rest of the time.

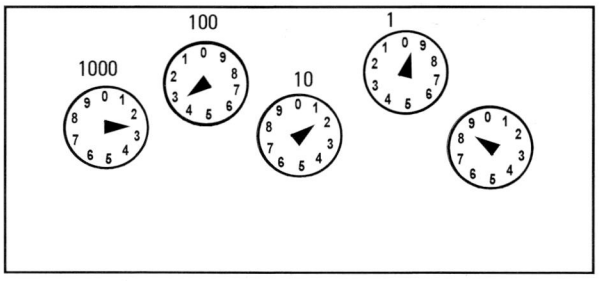

Figure 1.4

Meter reading

There are two methods of registering the units used by consumers. These are by
- analogue or dial meter
- digital or number meter

An **analogue meter** is one which consists of a number of dials with pointers to show the energy used.

These have to be read starting from the left and going across. For example, the reading for the meter shown in Figure 1.4 would be as follows.

On the 1000 dial the pointer is between 2 and 3. This means that more than 2000 units have been used, but not as many as 3000. So the reading is 2000 and something. From the 100 dial you can see that more than 300 units have been used but not 400. This means that the total so far is 23—. The 10 dial shows between 1 and 2, so is read as 10, making the total 231–. The final dial which is used for the reading is the 1, and this indicates between 0 and 9, making 9 the last numeral to record. The dial which records 1/10 (tenths) of units is disregarded for the purpose of meter reading.

Be careful about which way the numbering goes round the dials. You will see that they go one clockwise and the next anticlockwise across the five dials. The final reading for this meter is 2319 units.

The customer's reading for the previous quarter would be subtracted from this reading and the account worked out.

Try this

What is the reading for the meter shown below?

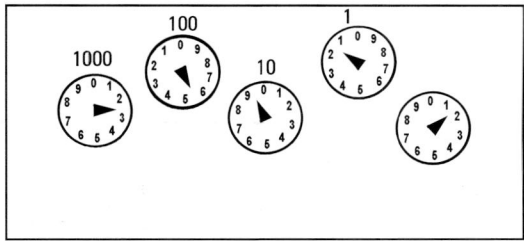

Reading _____

Draw arrows on the dials shown to indicate 5264 units used.

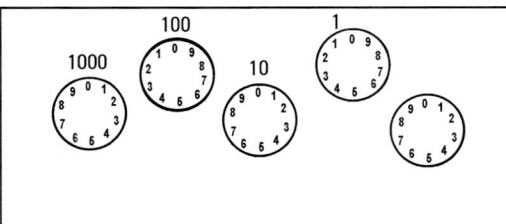

The digital meters are far easier to read for they display a series of numbers as shown in Figure 1.5. This leads to less confusion between the electricity boards and the customers.

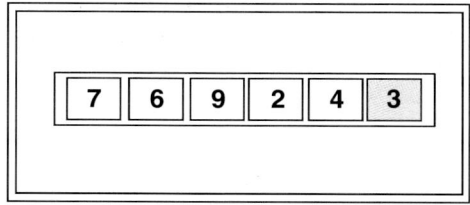

Figure 1.5

Where an Economy 7 tariff is being used a two scale digital meter is used, as shown in Figure 1.6. On the top scale all of the units used during the seven hours "cheap" rate are registered. The bottom scale indicates all of the other units used. The arrow between the scales shows which scale is being used at the time.

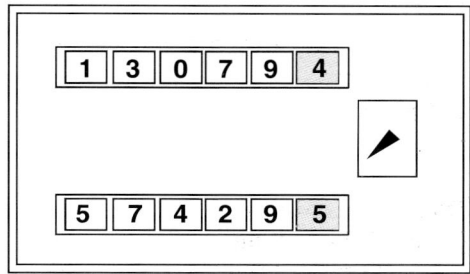

Figure 1.6

Examples using domestic tariffs

Remember there are two main tariffs in use.
These are
- Standard domestic rate
- Economy 7 domestic rate

Standard domestic rate

This is recommended if very little electricity is used at night.

Example

For each separate supply there would be a quarterly charge of

£8.35

AND

For each unit supplied   7.30p

This means that a household using an average of 7 units a day would use

7 × 7 days   = 49 units each week
and 49 × 13 weeks   = 637 units each quarter

637 units charged at 7.30p gives

$$637 \times 7.30 \qquad = 4650.1\text{p or } £46.50$$

plus a quarterly charge of £8.35

the total bill would be $\qquad = £54.85$

Try this

If a consumer uses an average of 8.5 units per day what would the quarterly bill be at the Standard domestic rate shown above?

Economy 7 domestic rate

This is used where there is an amount of electricity used at night. The meter used is the two scale type so that all units used during the seven hour night period are recorded separately.

Example

An example of this tariff could be

For each separate supply a quarterly charge of
£12.60

AND

For each unit supplied during seven hours between midnight and 8 a.m. Greenwich Mean Time (the actual times are fixed by the area electricity board)

2.45p

AND

For each unit supplied at other times
6.88p

Using the previous example, if 2 of the 7 units were used during the night period the cost would now be

	day units	night units
per day	5	2
per week	35	14
per quarter	455	182

cost per quarter	$455 \times 6.88\text{p}$	$182 \times 2.45\text{p}$
	= 3130.4p	= 445.9p
	or £31.30	or £4.46

total unit cost £31.30 + 4.46
$$= \qquad £35.76$$

quarterly charge £12.60
total bill £48.36

This shows a saving of £54.85 − £48.36
$$= £6.49 \text{ each quarter}$$
$$\text{or } £25.96 \text{ each year}$$

If there was no night usage of electricity but Economy 7 Domestic system was installed, the cost would be greater than using the standard domestic rate due to the higher quarterly charge.

VAT

All these costs are exclusive of VAT, which will be charged on the bill at the current rate.

Try this

A domestic consumer on Economy 7 tariff uses an average of 3.5 units during the economy 7 hours and a further 5.5 units during the day. What will the consumer's quarterly bill be if the rates are as shown in the example?

If the current rate of VAT at the time of the previous bill examples was 5% then the total for each bill would be:

The standard rate bill totalled £54.85.

£54.85 plus 5% VAT
$$= 54.85 + 2.74$$
$$= £57.59$$

The Economy 7 domestic tariff rate bill totalled £48.36.

£48.36 plus 5% VAT
$$= 48.36 + 2.42$$
$$= £50.78$$

Try this

A consumer uses an average of 10.2 units per day. What would the total quarterly bill be at the following Standard domestic rate?

A quarterly charge of   £8.35
AND

for each unit supplied   7.50p
Include VAT at 5%.

Try this

1. Get an up-to-date price leaflet from your Area Electricity Board for Domestic Tariffs.

2. Take the reading on a domestic meter.

   Reading

   Date

   Time

3. Take a second reading on the same domestic meter 24 hours after the first reading.

   Reading

   Time

4. Calculate the number of units used during the 24 hours.

   2nd reading

   1st reading

   Units used

5. Calculate the cost of units used over 24 hours.

   Cost per unit

   Units used

   Total cost

6. Add VAT at 5%

   TOTAL COST

Points to remember ◄ – – – – – – – – – – – – – –

The S.I. unit of energy is the _____

For electricity supply purposes the joule is too _____
_____

For convenience energy consumption for domestic consumers
is measured in _____

The electricity suppliers offer tariffs to encourage the use of
electricity during "off-peak" hours.

The cost of energy is made up of a standing charge and a price
per unit for the energy consumed. The standing charge covers
many costs which have to be met whether or not the consumer
uses electricity.

Two-rate energy meters controlled by teleswitches are used
for tariffs offering "economy rate" for energy used during the
_____

Self-assessment multi-choice questions
Circle the correct answers in the grid below.

1. A consumer uses an average of 6.25 units a week at 5.25p
   per unit. If a quarterly charge of £8 is made the quarterly
   bill would be
   (a) £40.81
   (b) £12.26
   (c) 931.25p
   (d) £37.85

2. If a domestic consumer on Economy 7 uses an average of
   5 units per night and 7 units each day what is the total
   consumption for the quarter?
   (a) 84 units
   (b) 1092 units
   (c) 156 units
   (d) 2898 units

3. 1 kWh expressed in joules would be
   (a) 3 600
   (b) 36 000
   (c) 360 000
   (d) 3 600 000

4. The cost of which of the following is not covered by the
   quarterly tariff
   (a) meter reading
   (b) billing and accounts
   (c) energy used
   (d) supplying service to cables, meters and the like

5. A consumer on the Standard rate tariff uses an average of
   11 units per day (3 of these units are used at night). What
   would the saving be if the consumer changed to Economy
   7 (excluding VAT)?
   The costs are as follows.
   On the Standard tariff the quarterly charge is £8.35 and
   the price per unit is 6.60p.
   On Economy 7 the quarterly charge is £12.60, the price
   per night unit is 2.45p and the price per day unit is 6.88p.
   (a) £5.04
   (b) £9.28
   (c) £13.74
   (d) £17.64

Answer grid

1	a	b	c	d
2	a	b	c	d
3	a	b	c	d
4	a	b	c	d
5	a	b	c	d

2

Cables

You will need to have a current copy of a cable manufacturer's catalogue available for reference in order to complete the exercises within this chapter. It could also be helpful to be able to refer to BS 7671.

Before you start this chapter obtain a copy of a manufacturer's catalogue or look at samples of cables in your local wholesaler but do *not* interfere with cables connected to the supply. If you are working with a college or training centre you could also find some samples to look at there. Give a typical stock no./catalogue no. for the following types of cable:

Rubber insulated flexible cable

Unshielded twisted pair (UTP)

PVC insulated and sheathed cable

On completion of this chapter you should be able to:

◆ identify the different parts that make up a cable
◆ explain why conductors are made of the materials they are
◆ select insulation types for given applications
◆ recognise the need to protect cables from mechanical damage
◆ recommend types of cable for given situations
◆ relate the diameter of a conductor to its cross-sectional area
◆ calculate the cross-sectional area of a conductor, given its diameter and number of strands
◆ compare solid and stranded conductors to their applications
◆ recognise the need for shaped conductors in large cables
◆ identify the factors to be considered when selecting cables
◆ select cables for given applications
◆ recognise the limitations of the materials used in the construction of some cables
◆ complete the revision exercise at the beginning of the following chapter

Part 1

Cable structure

Figure 2.1

The three main components

The job of a cable is to carry current. It must be able to do this safely and effectively without failing and without creating danger to persons, livestock or property in the process.

In order to carry current a cable must have an electrical conductor. This conductor must be of sufficient size to carry its predicted load without overheating and thus causing other problems.

The conductors have to be insulated from each other and from any surrounding metal surfaces. This ensures that the current flows along its predetermined path and also prevents the possibility of electrical shock if anyone should happen to touch the cable or its enclosure.

The insulation will only remain effective if it remains undamaged or is protected against contamination. This means that in most cases, the insulated conductor has to be provided with some form of mechanical protection in the form of covering, sheathing or enclosure.

Looking back over what has been said you will see that a cable used as a live conductor can be divided into three basic parts (Figure 2.2), namely:

- the conductor
- the insulation
- the mechanical protection

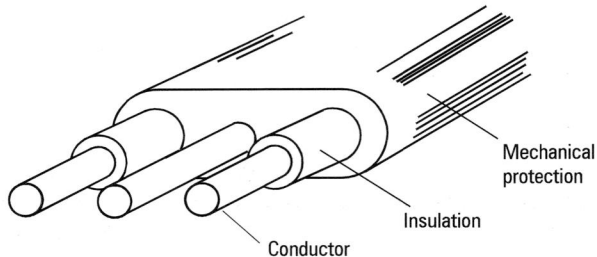

Figure 2.2   *The three main components of any cable*

Now we'll take a look at each one in turn.

The conductor

First of all, the conductor needs to be a good conductor of electricity, or in other words, it needs to have a low resistivity.

We could choose silver for this purpose as it is an excellent conductor at normal temperatures, but sheer economics rules it out as a possibility.

Very closely behind silver comes copper. It is a very good conductor, it has good mechanical properties, it is flexible and it is easy to terminate. Copper may not be cheap, but it's a lot cheaper than silver and years of experience have shown that it has a lot going for it.

If you are looking at the pure economics of cable manufacture you really must give aluminium some serious consideration.

It may not be as good a conductor as copper, having a resistivity about 1.6 times greater (see Figure 2.3), but a look at the commodity prices in your morning paper will show that you could well afford to make a bigger cable and still save money.

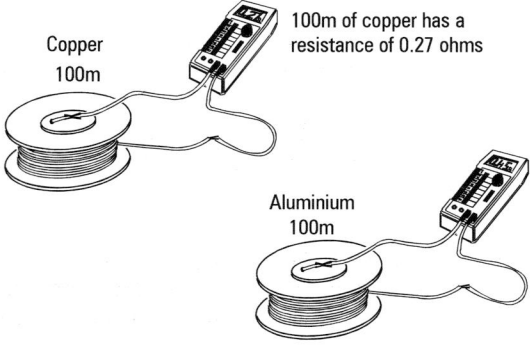

Copper 100m

100m of copper has a resistance of 0.27 ohms

Aluminium 100m

Figure 2.3

100m of the same sized aluminium cable will have a resistance of 0.45 ohms

Another reason why you might choose aluminium is the fact that it is much lighter than copper; the density of aluminium being less than a third of that for copper (see Figure 2.4). Those two facts alone must make aluminium a strong contender in the conductor stakes.

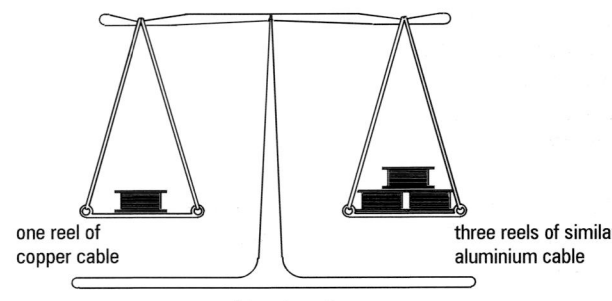

one reel of copper cable

weighs about the same as

three reels of similar aluminium cable

Figure 2.4

There are plenty of other conductors about, such as iron, lead, tin, brass and zinc, but they all have their problems, so that when we're thinking of cables we keep coming back to the same two:

- copper or
- aluminium

Before deciding on aluminium though, consider this: Aluminium is difficult to terminate as it oxidises easily and is prone to electrolytic corrosion.

Bearing that in mind, the choice (Figure 2.5) could be up to you.

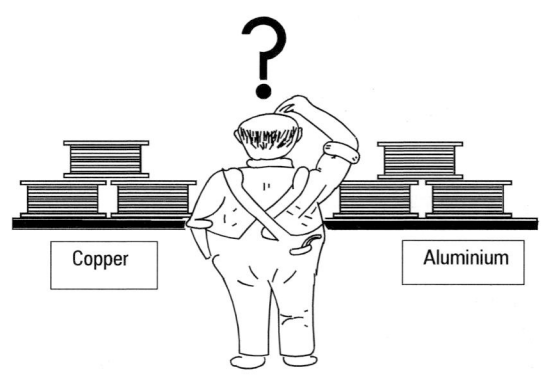

Copper   Aluminium

Figure 2.5

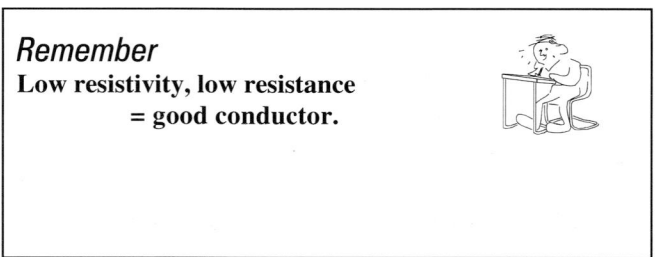

Remember
**Low resistivity, low resistance
= good conductor.**

The insulation

If you've any experience of cables you'll probably think of insulation as **PVC**, and that's it.

You'd be quite justified in thinking that PVC (**polyvinyl chloride**) is by far the most common form of cable insulation in use today and you've really got to hunt around to find anything different. However, XLPE (cross-linked polyethylene) and LSF (low smoke and fume) cables have grown in popularity for many types of installation because of their electrical, thermal and environmental properties.

The insulation has to keep the current in the conductor, so its most important feature is that it must have a high electrical resistance.

Air is a good insulator and this is the reason why overhead transmission is carried out using bare conductors (Figure 2.6) separated from each other and from earth by the air, which insulates and cools them. Where they are supported by towers or poles these bare conductors are suspended from glass or porcelain insulators which are mechanically strong and have the very high resistivity necessary for insulation purposes.

Being a good insulator is not the whole story though.

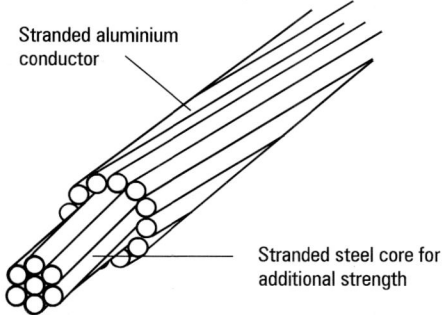

Figure 2.6   Bare overhead line conductor (air-insulated)

A **PVC** insulator (Figure 2.7) has to do its job effectively in some pretty arduous conditions. PVC may be cheap, flexible, easily coloured and impervious to water, but it tends to soften as the temperature increases above 70 °C, and at temperatures below 0 °C it may become brittle and crack when handled.

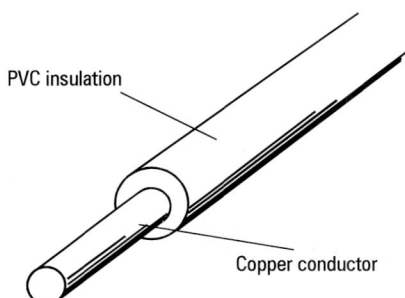

Figure 2.7   Single-core PVC-insulated copper cable

Butyl rubber is a synthetic rubber which is frequently used for high-temperature terminations such as heating elements. It will operate continuously without deterioration in temperatures up to 85 °C.

Silicon rubber insulation can be used on cables in situations where temperatures are very high. It has an operating temperature of 150 °C and even if the cable is completely burnt, the insulation forms a silicon ash which is in itself an insulator, so the cable may continue to function until the ash is disturbed.

For even higher temperatures (up to 185 °C) glass fibre insulation may be used, and you can frequently find examples of this type of insulation on the internal wiring of heating appliances.

> *Remember*
> **Conductors**
> must have a low resistivity
>
> **Insulators**
> must have a high resistivity

MIMS (Mineral-insulated metal-sheathed)

One of the most remarkable insulating materials is **magnesium oxide** (or **magnesia**). This is a white powder not unlike powdered chalk to look at.

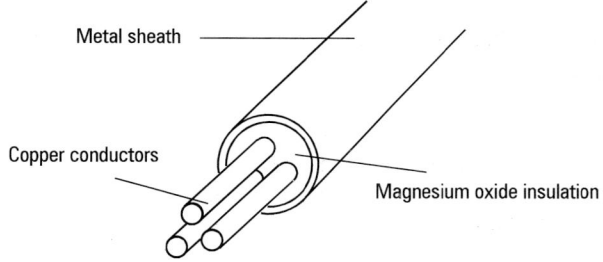

Figure 2.8   Mineral-insulated metal-sheathed cable

Magnesium oxide is virtually indestructible and keeps its insulating properties even though its temperature rises to well above 1000 °C, after which the conductors will have melted and broken down. This is the insulating material which is to be found in Mineral-Insulated Cable, a type of cable renowned for being robust (Figure 2.8).

Magnesium oxide must however be kept completely dry, and since it can absorb moisture from the atmosphere, all terminations must incorporate a seal to act as a moisture barrier.

Paper insulation

In some cables, notably lead-sheathed armoured cable, **paper** is used as an insulator (Figure 2.9). For this purpose, paper which has been impregnated with mineral oil is applied to the conductor in the form of a multi-layer tape. Impregnated paper is an excellent insulator and is to be found in cables operating at very high voltages.

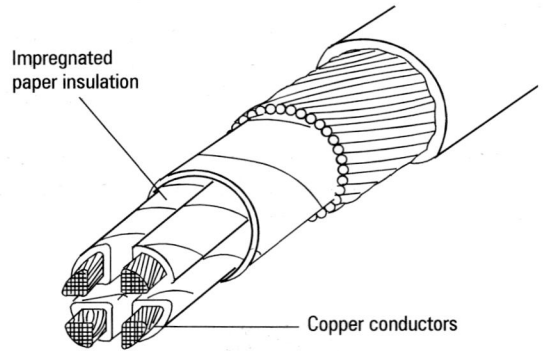

Impregnated paper insulation

Copper conductors

Figure 2.9   Paper-insulated steel wire armoured cable

As with magnesia, the paper insulation can be broken down if water penetrates the cable, and for this reason paper-insulated cables must have a watertight sheath and sealed terminations.

Low toxicity

Before leaving the topic of insulation we must consider the effects of smoke emission. General-purpose PVC cable, when it becomes very hot, can emit poisonous smoke, which can create a serious hazard. This danger is well known, and many contracts now carry a stipulation that low-toxicity (LSF) cables have to be used.

Points to remember ◄ – – – – – – – – – – – – – – – –

We have seen that all cables consist of conductors, some form of _____ and a method of _____ from mechanical damage.

The conductors generally are made of _____ or _____, depending on the installation requirements.

Nowadays, the insulation of the conductors is often PVC, but could be one of the other materials discussed. List those that you can you remember, then look back and check your answers.

To protect the conductors and their insulation from mechanical damage, an outer layer is used around the cable. The nature of this layer will depend on the degree of mechanical protection required.

All cables should be chosen taking into account the limitations of the materials used compared with the environment in which they are to be installed.

Try this

As a personal project find out what other cable manufacturers have to offer in the way of low-toxicity cables and list them below.

Manufacturer	Cable name	Catalogue number

Part 2

The mechanical protection

Bare conductors are supplied for overhead transmission, use air for insulation, and require no further mechanical protection. Insulated conductors are available to the installer who is going to make his or her own provision for the mechanical protection of the cables.

Cables in conduit and trunking

This type of cable is usually in the form of PVC-insulated singles, as shown in Figure 2.10. These have a conductor surrounded by a single layer of PVC insulation and no further covering. PVC singles are normally installed in conduit or trunking or as part of the internal wiring of apparatus.

Being surrounded by a protective enclosure, they are not exposed to the risk of mechanical damage and it is normally quite unnecessary to make any further provision.

Do, however, make sure that the cable is not damaged during installation by drawing too many through a small aperture or by leaving sharp metal edges unprotected.

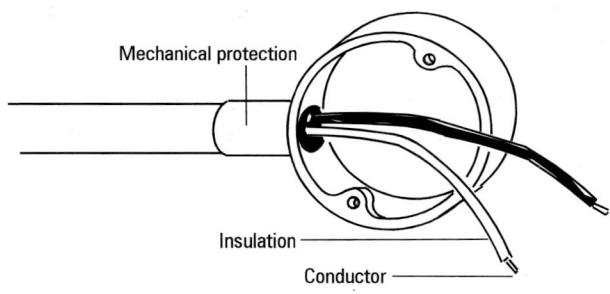

Figure 2.10   *PVC-insulated single-core cable in conduit*

PVC-insulated and sheathed cable (Figure 2.11) is used a great deal for domestic and light commercial wiring. This has PVC-insulated conductors which are surrounded by a second overall sheath of PVC.

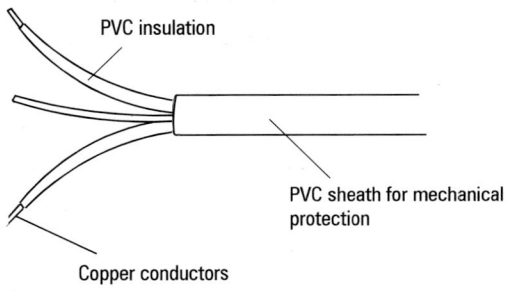

Figure 2.11   *PVC insulated and sheathed cable*

This type of cable can be used where the risk of mechanical damage is slight. It is normally run above ceilings, under floors and within walls and often needs no further protection, although additional protection can be given where necessary.

Armoured cable (Figure 2.12), in addition to insulation and sheathing, incorporates a layer of metallic armouring which can protect the cable from impact.

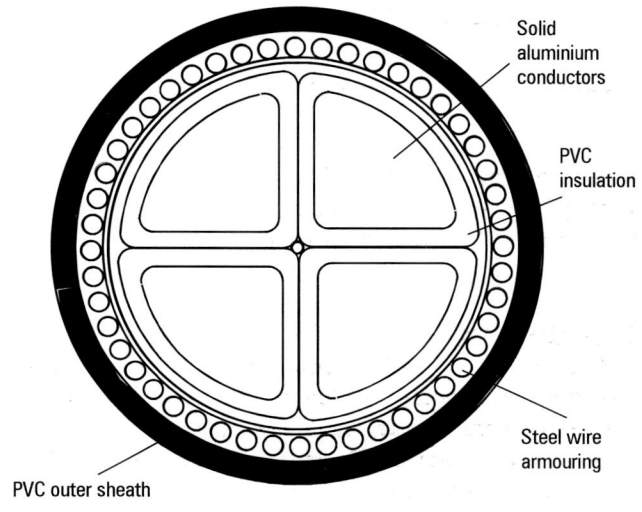

Solid aluminium conductors

PVC insulation

Steel wire armouring

PVC outer sheath

Figure 2.12

This type of cable is widely used in industrial installations, where it is to be seen clipped directly to the surface or run in ducts or trenches.

MIMS, mineral-insulated metal-sheathed cable, has a seamless metallic (normally copper) sheath and solid conductors (Figure 2.13).

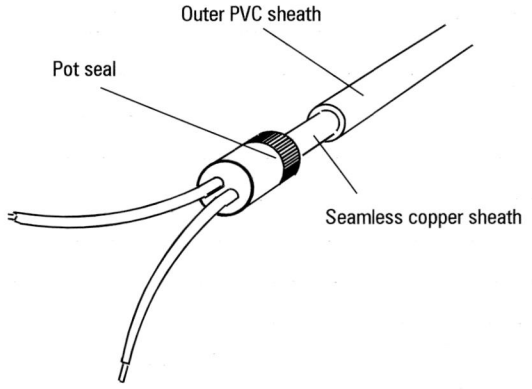

Outer PVC sheath

Pot seal

Seamless copper sheath

Figure 2.13   *Mineral-insulated metal-sheathed cable with an overall covering of PVC*

The insulating medium is magnesium oxide, as previously described, and this is a very robust cable. MIMS cable can put up with some very rough treatment and can withstand temperatures far in excess of those which would destroy any other cable.

Additional protection from the effects of electrolytic corrosion when in contact with other metals in damp surroundings can be given by means of a PVC outer sheath.

Remember

If the metallic sheath is broken in any way the insulation will quickly fail due to the penetration of moisture.

PX and FP 200 coaxial cables are of similar construction and can sometimes be an alternative to MIMS where conditions are less arduous (Figure 2.14).

The insulated conductors are enclosed in an aluminium sheath and the whole cable is enclosed in an outer PVC covering.

This type of cable can be concealed in the fabric of a building or clipped directly to the surface. They are resistant to a fair amount of mechanical damage and can also be used in situations where conditions are wet or dirty.

When using PX and FP200 cable, manufacturers' data must be consulted for current ratings and resistance.

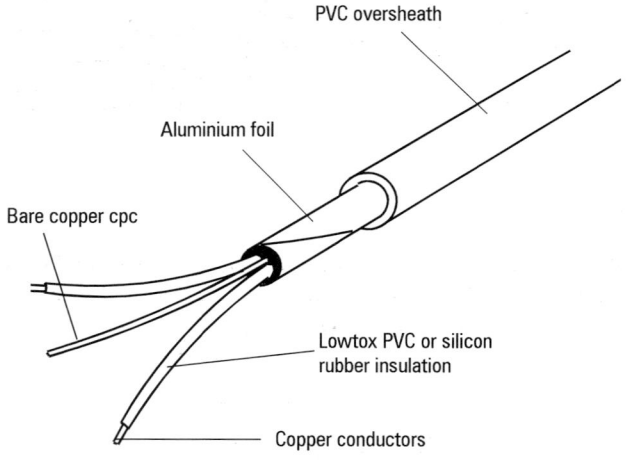

PVC oversheath

Aluminium foil

Bare copper cpc

Lowtox PVC or silicon rubber insulation

Copper conductors

Figure 2.14   *PX and FP 200 aluminium-sheathed coaxial cable*

Quiz

You have now looked at several words associated with electrical cables. The words in the list below are included in the square. Mark each word as you find it. They may be found in any direction.

F	M	U	I	N	I	M	U	L	A
P	R	O	T	E	C	T	I	O	N
A	E	I	H	V	O	V	C	H	A
P	A	T	A	L	N	O	P	S	I
E	R	I	W	I	D	E	N	M	S
R	O	T	A	L	U	S	N	I	E
L	E	L	B	A	C	U	R	M	N
I	S	H	E	A	T	H	T	C	G
O	D	E	R	U	O	M	R	A	A
C	O	P	P	E	R	O	C	O	M

CABLE	PVC
SHEATH	ARMOURED
MIMS	PROTECTION
COPPER	PAPER
ALUMINIUM	AIR
CONDUCTOR	CORE
INSULATOR	WIRE
MAGNESIA	COIL

The job of the conductor

Now that we've looked at cables in general, let's take a look at the conductor part in some more detail.

The job of the conductor is to carry current from the supply and deliver it to some other part of the installation without getting too hot and bothered and without losing too many volts in the process.

Figure 2.15

Choosing the right conductor (Figure 2.15) is about the most important job the designer or installing engineer has to do.

Getting it right means that the job progresses as planned and everybody is happy.

Getting it wrong means delay, expense and moans all round.

Selecting the conductor

When you choose a conductor to do a job it's got to be big enough. This is not just a desirable feature, it is in fact a regulation and, in most cases, a statutory requirement. On the other hand, if you go over the top you'll soon be accused of wasting somebody's money and this will do your reputation no good at all.

Let's get it right.

Let's get back to basics so we all know what we are talking about – let's look at the whys and wherefores of conductors.

Solid conductors

Small conductors for fixed wiring are likely to have single solid cores of fairly small diameter, as shown in Figure 2.16.

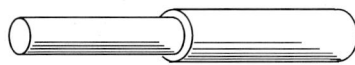

Figure 2.16

These can easily be bent to go round bends or form terminations. Figures 2.17 and 2.18 show examples.

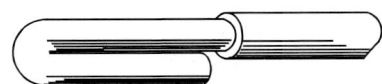

Figure 2.17

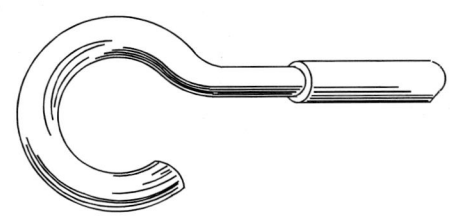

Figure 2.18

Stranded conductors

As conductor sizes get bigger you will find that the core will consist of several strands of wire.

This makes the cable more flexible and easier to work with.

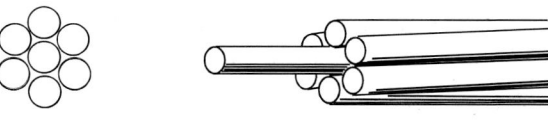

Figure 2.19

A seven-stranded conductor will consist of six outer strands which have been laid in a spiral around a single central strand (Figure 2.19). All the strands have the same diameter and they all go together to make up a single conductor core.

As cables get larger, even seven-stranded conductors could prove to be too stiff and inflexible. For easy handling the conductor core will be made up of 19 strands.

A 19-strand conductor will be made up of a single strand surrounded by six others as before, but now surrounded by 12 strands which have been twisted in the opposite direction to those underneath (Figure 2.20).

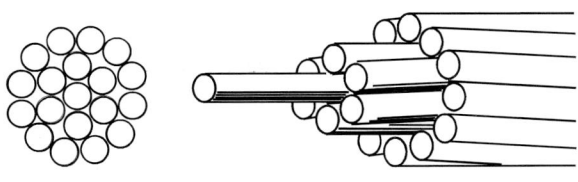

Figure 2.20

Taking the same idea even further you will find that conductors having 37 strands, laid up in successive layers, go to make up the larger sizes of conductor (Figure 2.21).

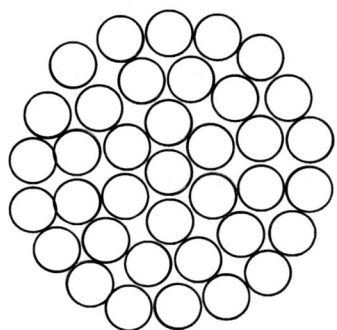

Figure 2.21   *37 core cable*

Shaped conductors

In order to make round conductors fit inside a round overall sheath the manufacturer has to add some form of filling to take up the spaces between the cores of multi-core cables (Figure 2.22). With very large cables, the amount of filling could be quite considerable and would make the overall diameter of the cable much larger than it needed to be.

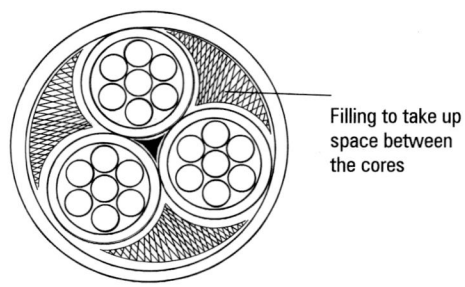

Filling to take up space between the cores

Figure 2.22

This problem can be overcome by using shaped conductors which have been rolled to form a wedge shape not unlike that of a familiar type of spreading cheese (Figure 2.23).

Figure 2.23

This makes the cable much more compact and no doubt helps in the manufacturing process.

Examples of shaped conductors are shown in Figure 2.24.

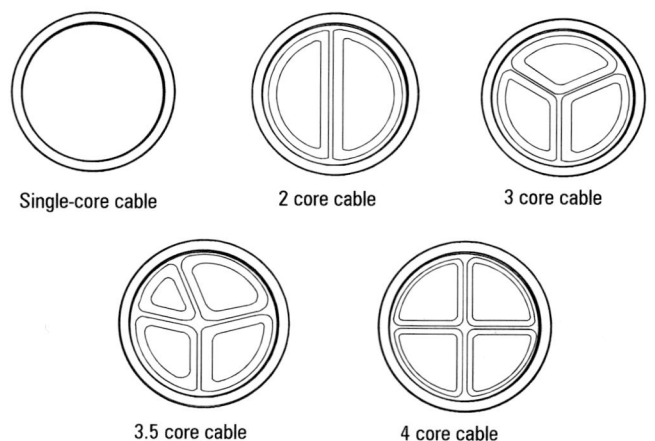

Single-core cable   2 core cable   3 core cable

3.5 core cable   4 core cable

Figure 2.24

Good conductors, like copper and aluminium, offer very little resistance to current flow under normal conditions but there are several factors that can change this.
They are:

- the length of the conductor
- the cross-sectional area of the conductor
- the temperature of the conductor and of course
- the material the conductor is made of (called the resistivity of the material).

Points to remember ◄ – – – – – – – – – – – – –

The three principal components of a cable are conductor, electrical insulation and mechanical p_____

Conductors have a low resistivity and insulators have a high resistivity. The most common conductors in use are c_____ and a_____.

Multicore cables comprise a number of conductors, with electrical insulation as appropriate, contained within a single mechanical protection sheath.

Bare conductors need to be installed in such a way that they do not come into contact with anything else, or each other.

Insulated and sheathed cables have their own mechanical protection and are suitable for general wiring. Conductors with electrical insulation only must be installed in a containment system, such as conduit or trunking, to prevent damage to the i_____

Cables with integral armouring beneath the cable sheath are used for installations in adverse conditions, such as underground, where conduit and trunking are not appropriate.

Try this

Using a manufacturer's catalogue find out the listed numbers for the cables below.

2 core PVC insulated and sheathed

3 core paper insulated

4 core PVC insulated steel wire armoured

Self-assessment multi-choice questions.

Circle the correct answer in the grid below.

1. Which of the following is the best conductor?
   - (a) copper
   - (b) aluminium
   - (c) iron
   - (d) silver
2. Copper may be used in preference to aluminium as a cable conductor because it is
   - (a) cheaper
   - (b) lighter
   - (c) a better conductor
   - (d) used for cables greater than 16 mm^2
3. Aluminium may be used in preference to copper because it
   - (a) has less tendency to corrode
   - (b) is lighter
   - (c) is easier to terminate
   - (d) is a better conductor
4. Which of the following is the most widely used cable insulator?
   - (a) PVC
   - (b) butyl rubber
   - (c) silicon rubber
   - (d) magnesium oxide
5. Which of the following insulators can withstand the highest temperature?
   - (a) magnesium oxide
   - (b) glass fibre
   - (c) silicon rubber
   - (d) PVC
6. Overhead line conductors are insulated from each other and earth by
   - (a) PVC
   - (b) rubber
   - (c) magnesia
   - (d) air
7. Which of the following insulators emits toxic gas when heated?
   - (a) magnesium oxide
   - (b) PVC
   - (c) paper
   - (d) glass fibre
8. What form of sheathing is normally applied to cables for domestic and light commercial wiring?
   - (a) lead
   - (b) copper
   - (c) PVC
   - (d) steel wire
9. The purpose of the seals on the terminations of MIMS cable is to
   - (a) make sure that the sleeves do not become detached
   - (b) insulate the termination from the sheath
   - (c) keep the magnesia from escaping
   - (d) prevent moisture from entering the cable

10. The type of cable shown in Figure 2.25 is

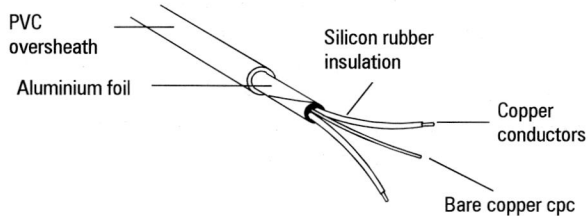

Figure 2.25

   - (a) PVC-insulated and sheathed
   - (b) MIMS
   - (c) FP 200
   - (d) PVC insulated steel wire armoured

Answer grid

1	a	b	c	d		6	a	b	c	d
2	a	b	c	d		7	a	b	c	d
3	a	b	c	d		8	a	b	c	d
4	a	b	c	d		9	a	b	c	d
5	a	b	c	d		10	a	b	c	d

Part 3

Calculation of conductor cross-sectional areas

When you've got an electrical load to supply you need a conductor to carry the current to the load. In the simplest possible terms; the bigger the current – the bigger the cable.

So, how big is big?

A round conductor, looked at end on has a circular shape (Figure 2.26).

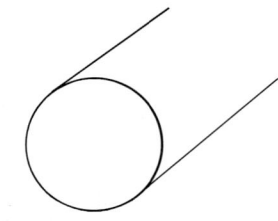

Figure 2.26   Single-core cable

As you probably know, a circle has a diameter. This is the distance from one side to the other in a straight line through the centre (Figure 2.27).

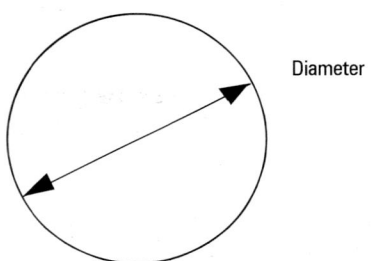

Diameter

Figure 2.27

Is this the "size" of the conductor?

No, not quite.

Let's take a simple example (Figure 2.28).

Cable A is 1 mm in diameter.

Cable B is 2 mm in diameter.

You could put two A's inside one B and there's still a lot of conductor left over.

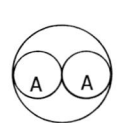

Figure 2.28

So how do we do it?

Cross-sectional area, that's the answer.

Remember

The cross-sectional area can be calculated from the formula:

$$\text{c.s.a.} = \frac{\pi \times d^2}{4}$$

Note

d is the diameter of the conductor.

Example

We will use the conductors in Figure 2.28 and measure their diameter, using a Vernier gauge, as shown in Figure 2.29.

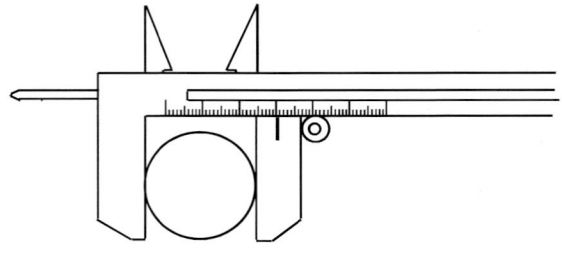

Figure 2.29

	A	**B**
Take the diameter	1 mm	2 mm
Square it (multiply it by itself)	1	4
Multiply it by π (3.1416)	3.1416	12.566
Divide the result by 4	0.7854 mm^2	3.1416 mm^2

So if the diameter of one conductor is twice the size of another then its cross-sectional area will be four times as great.

Of course a multi-strand conductor will have a cross-sectional area equivalent to that of one strand (Figure 2.30) multiplied by the number of strands.

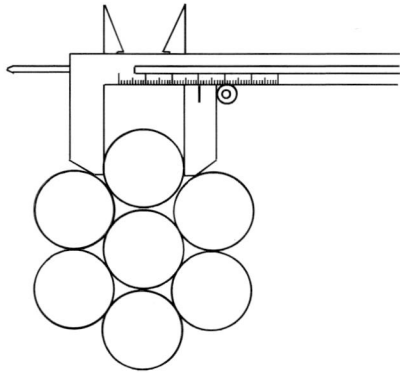

Figure 2.30

Example

A 10 mm² conductor has 7 strands each measuring 1.35 mm in diameter. The total cross-sectional area will be:

each strand

$$\frac{1.35^2 \times \pi}{4} = 1.4314 \text{ mm}^2$$

total

$$1.4314 \times 7 = 10 \text{ mm}^2$$

If you look at a list of cable data in a manufacturer's catalogue you will see that all the conductors are listed under their cross-sectional areas.

Try this

Calculate the cross-sectional area of each of the following:

7 strands of 2.14 mm diameter

19 strands of 1.78 mm diameter

37 strands of 2.25 mm diameter

61 strands of 2.85 mm diameter

Flexibility of conductors

Flexibility is an important feature of any cable. Some cables are used to provide flexible supplies to moving machinery (Figure 2.31). These are sometimes known as trailing cables!

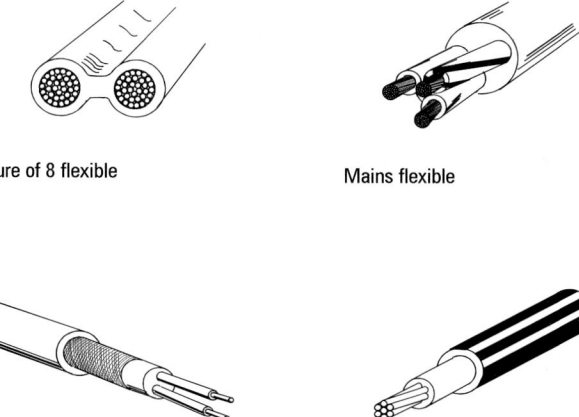

Figure of 8 flexible

Mains flexible

Screened flexible

Heavy duty ethylene propylene rubber (EPR)

Figure 2.31   Flexible cables

The cores of these cables are made up of large numbers of strands, each strand having a very small diameter.

For example, a small cable for very light duty could have 16 strands 0.2 mm in diameter making a cross-sectional area of 0.5 mm².

At the other end of the range the same type of cable could have 485 strands, each of 0.5 mm in diameter making a cross-sectional area of 95 mm².

One manufacturer lists 400 mm² cable with 2013 strands of 0.5 mm. This is a very big cable indeed, so don't get the idea that flexibles come in small sizes only.

Rigid conductors

Of course, flexibility isn't always necessary, and some cables, once laid or fixed in position, do not need to be moved. A good example of this is solid aluminium underground cable which has solid wedge-shaped conductors (Figure 2.32). This is not very flexible, but flexible enough to wind off a drum and lay in a trench or duct.

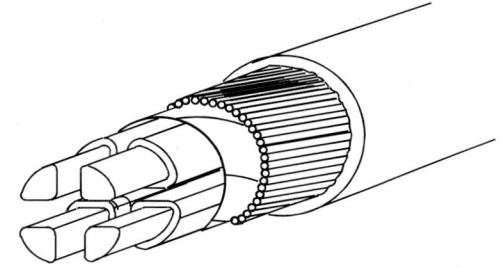

Figure 2.32   Solid aluminium conductors

Current-carrying capacity of conductors

For all stranded cables from 1 mm^2 to 400 mm^2 you will find at least one, but most probably several, tables of current rating.

It's no problem reading a table on a straightforward one-to-one basis, and this can give you some idea of how much current a given cable can carry.

As ever, there's more to it than that, and later we'll be looking at the process of cable selection in some more detail, but let's just take some examples. Using Table 4D2A from BS 7671 (The IEE Wiring Regulations) or a cable manufacturer's catalogue, enter the appropriate current ratings in the "Try this" below.

You will see that the current-carrying capacity does not go up in direct proportion to the cable size, so don't expect a cable twice the size to carry twice the current – it won't.

Try this

Complete the following table using information from either BS 7671 Table 4D2A or a cable manufacturer's catalogue.

Table 2.1 Current-carrying capacity

Standard Copper Conductors 1 three-core core cable, PVC-insulated, non-armoured in accordance with BS 6346/1969.			
Conductor cross-sectional area	Cable buried directly in ground	Cable run in a duct	Cable clipped in free air
mm^2	A	A	A
1			
1.5			
2.5			
4			
6			
10			
16			
25			
35			
50			
70			
95			
120			
150			

A 30 A circuit would need a cable rated at 30 A or greater.
A 60 A circuit would need a cable rated at 60 A or greater.
An 80 A circuit would need a cable rated at 80 A or greater.

Although a 16 mm^2 cable has a c.s.a. (cross-sectional area) four times that of a 4 mm^2 cable, it is not capable of carrying a current four times as great.

Aluminium conductors

Because aluminium is not as good a conductor as copper, you will find the current-carrying capacity is less in each case.

Try this

Complete the following table using information from BS 7671, Table 4D2A, column 6, and Table 4K2A, column 6, or from a cable manufacturer's catalogue.

Table 2.2

Current-carrying capacity of PVC non-armoured cables

c.s.a. (mm)^2	Copper (A)	Aluminium (A)
16		
25		
35		
50		

Conductor selection

The nominal current of a circuit is that of the fuse or circuit breaker protecting it.

Example

A 6.5 kW load may have a design current of 27 A. The fuse protecting this circuit would have to be 30 A or 32 A, depending on the type, because no 27 A fuse is available. The cable would have to be rated at a current no less than that of the fuse, otherwise it would not be protected in the event of an overload.

Therefore, using the details you have listed in Table 2.1, we would go for 4 mm^2 cable rather than 2.5 mm^2.

A fuse does not blow at its rated current, but can carry a moderate overload for a period of time before operating.

The current which causes effective operation of the fuse must never be greater than 1.45 times the current rating of the cable it protects.

With some fuses, notably semi-enclosed rewireable types, the current-carrying capacity of the conductors must be reduced, otherwise the fusing current would be too high for the cable installed.

Cables are made with either solid or _____ conductors. The type of conductor used should be considered together with its application.

Cables are listed by their cross-sectional areas, but when round conductors are measured it is the _____ that is found. It is therefore necessary to carry out a calculation to convert this to the cross-sectional area. Where the conductor is made of more than one strand each one must be taken into account when calculating the total cross-sectional area.

Where large conductors are used to make up a cable consisting of more than one core, the conductors are _____ to keep the overall shape of the cable round. When comparing the current carrying capacity of cables the material that the conductors are made of must be considered together with the cross-sectional area.

Try this

Larger sizes of multi-core cables have shaped conductors rather than round ones. The purpose of this is to

A cable having 19 strands each with a diameter of 1.53 mm has a nominal c.s.a. of

A 70 mm^2 cable has 19 strands each with a diameter of

If cable A has half the diameter of cable B, which of the following statements is true?
(a) cable B has twice the c.s.a. of cable A
(b) cable B has four times the c.s.a. of cable A
(c) cable B has twice the current-carrying capacity of cable A
(d) cable B has four times the current-carrying capacity of cable A

Part 4

The environment

Before choosing a cable for a particular job the person responsible for the design should be fully aware of any environmental hazards which are likely to be encountered during the life of the installation.

Some of the things which could give you problems are:
- the risk of mechanical damage
- extremes of temperature – hot or cold
- the presence of moisture
- the presence of corrosive substances or other chemical hazards
- the likelihood of damage from domestic animals or vermin
- fire or explosion risks
- vandalism

As you probably realise, this is not a complete list, but you now have some idea of the problems you are likely to come up against when making your choice.

Protection is the name of the game, but the level of protection should always be appropriate to the risk involved. Other factors must always be taken into consideration. These are:
- economics
- appearance
- ease of installation

Types of cable

We have already looked at some of the types available. We'll go through them again and see how they would stand up to some of the environmental problems.

PVC singles REF 6491X

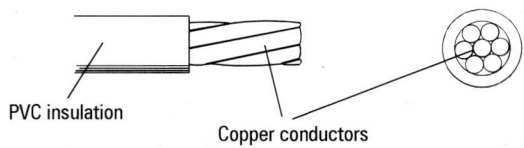

PVC insulation   Copper conductors

Figure 2.33

This cable (Figure 2.33) has no mechanical protection of its own, but can be used in conduit or trunking. This combination makes an excellent wiring system.

PVC singles in steel conduit is the type of installation you are likely to find in factories, workshops, warehouses, multi-storey car parks and so on.

Where the atmosphere is damp or where the installation is exposed to the weather, the conduit and its fittings may be galvanised in order to resist corrosion.

Where mechanical damage is less of a problem but moisture or the presence of other corrosives is going to give trouble, PVC conduit may prove to be a better alternative. The combination of PVC single cables in PVC conduit (Figure 2.34) is popular in agricultural and horticultural premises.

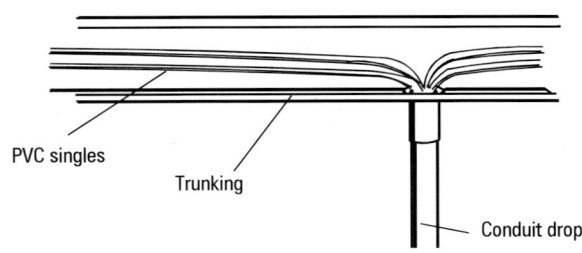

Figure 2.34

PVC sheathed and insulated – twin and multicore REF 6242 & 6243Y

This is perhaps the most popular system of all (Figure 2.35). The vast majority of domestic dwellings are wired in what is commonly referred to as "PVC twin and earth". Indeed, a considerable amount of this kind of cable is to be found in shop and office wiring as well.

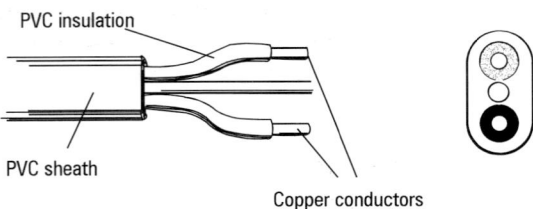

Figure 2.35

Mechanical protection is provided by the outer PVC sheath, and the circuit protective conductor ensures that there is good earth continuity to all outlets. The degree of mechanical protection is not high, but it is quite sufficient to prevent damage to the insulation during the installation process and since it is normally concealed in the structure of the building it is not normally subject to mechanical damage.

When PVC-sheathed cable is run on the surface it is normal practice to keep it out of harm's way, but where your peace of mind is at stake the cable can be drawn into conduit or mini-trunking to give it the extra protection needed (Figure 2.36).

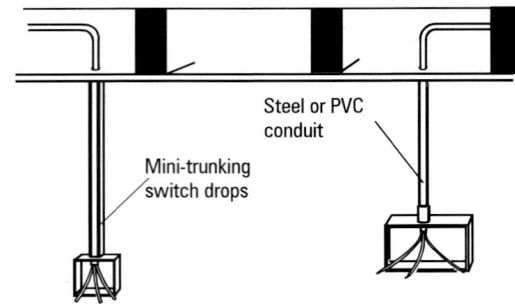

Figure 2.36

Mineral-insulated metal-sheathed cable BS 6207

MIMS cable (Figures 2.37 and 2.38) is very tough. It can withstand very high temperatures and can also put up with a great deal of ill treatment in the form of knocks and bangs. The construction of the cable is basically similar for all types, but it is generally listed as "heavy duty" or "light duty" with or without PVC oversheath.

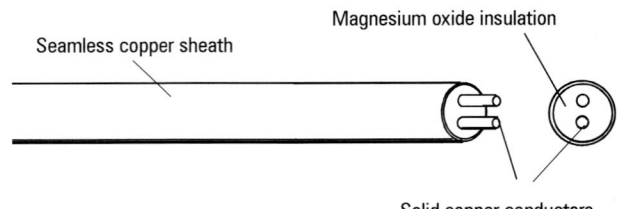

Figure 2.37

It can have as many as 19 cores, which may range in cross-sectional area from 1 mm^2 up to 150 mm^2.

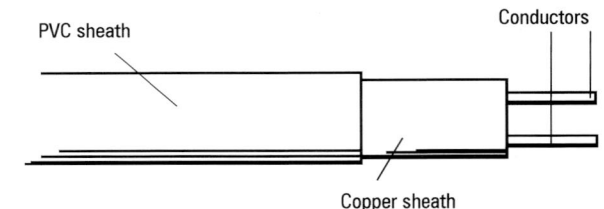

Figure 2.38

Mineral-insulated cable with an aluminium sheath is also available.

Applications suitable for **MIMS** cable are to be found in a wide range of installations. Although it is generally considered to be expensive for domestic wiring, you will find it used in factory workshops, churches, and in the dairy and food processing industries. It is frequently recommended for fire alarms and is sometimes the only type of cable suitable where environmental conditions are very severe.

The metal sheath of the cable is used as the circuit protective conductor, and if it is to be used with insulated fittings must have an "earth tail" incorporated in the termination.

Silicon rubber insulated FP200 cable

This is a very versatile cable (Figure 2.39). Its overall PVC sheath makes it suitable for damp or corrosive atmospheres and its aluminium tape acts as a metallic screen.

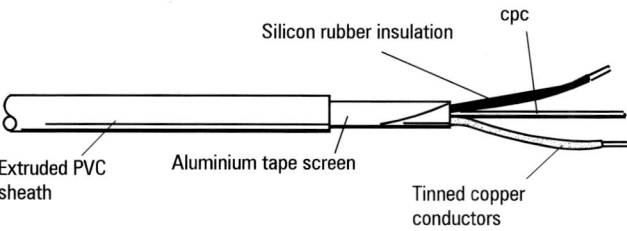

Figure 2.39

The silicon rubber insulation is very soft and prone to damage, and must be handled with care during termination, but within the cable it is adequately protected. The most remarkable feature of silicon rubber is its ability to act as an insulator even after the cable has been destroyed by fire. The insulation, after it has been burnt, turns into silica ash which continues to keep the conductors apart, and this helps to maintain essential services such as fire alarms and emergency lighting long after PVC insulation would have given up.

Try this

On the list below write the method of mechanical protection incorporated in each cable.

PVC twin and cpc cable

FP200 cable

mineral-insulated cable

Armoured cable

PVC steel wire armoured BS 6346

Being steel wire-armoured, these cables (Figure 2.40) are suitable for external use in ducts or buried directly in the ground. They are also sufficiently well protected to be clipped directly to the surface inside buildings, even where there is some risk of mechanical damage.

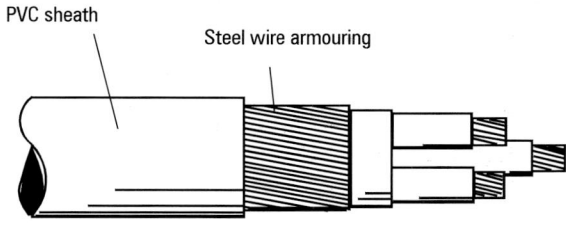

Figure 2.40   PVC steel wire-armoured BS 6346 cable

This type of cable is widely used in industry, where it is popular because the outer sheath gives the cable protection against moisture and other potentially harmful substances. The steel wire armouring provides the necessary mechanical protection, may act as a circuit protective conductor and is flexible enough to permit easy handling of the cable.

As well as PVC, a material usually referred to as XLPE is now very popular as an insulating material. XLPE, or cross-linked polyethylene, to give it its full title, has a much wider operating temperature range than PVC because it retains its mechanical strength at high temperatures long after PVC has gone soft. Therefore XLPE-insulated cables can carry more current than PVC-insulated cables for the same c.s.a. of cable.

Aluminium tape armoured cable

Aluminium is widely used in armoured cables, PVC and XLPE, both as an armouring layer and as a conductor. Aluminium-armoured cable (Figures 2.41 and 2.42) does not have the same current-carrying capacity as copper, but is very much lighter, and this can prove to be something of an advantage during installation.

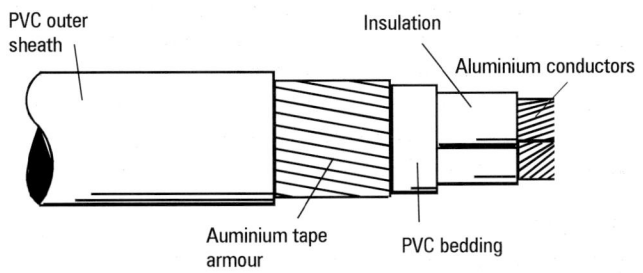

Figure 2.41   Aluminium tape armoured cable

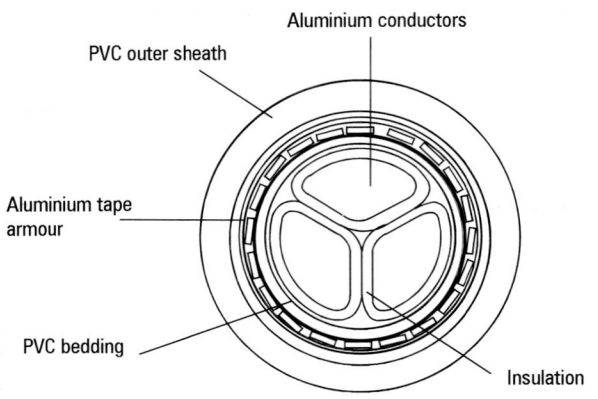

Aluminium conductors

PVC outer sheath

Aluminium tape armour

PVC bedding

Insulation

Figure 2.42   Structure of aluminium tape-armoured cable

Before leaving PVC insulation to consider other types, it must be noted that it is a requirement of BS 7671 that cables which are susceptible to damage at low temperatures should be treated with special care.

PVC-insulated or sheathed cables should preferably not be installed where the temperature is consistently below 0 °C, and if the cable has been kept at a temperature which is below freezing, then it should not be handled or installed until it has had sufficient time to reach a reasonable handling temperature.

Try this

Obtain a selection of cables in as many different sizes as you can. (For example 1.0 mm², 1.5 mm², 2.5 mm², 4.0 mm², 6.0 mm², 10 mm² and 16 mm²).

Practice identifying the size of a cable by looking at it.

First compare the different sizes and grade them. Then identify each one by its cross-sectional area.

Make a note below of all the different sizes you have identified.

 Take the necessary safety precautions when using blowlamps.

Using MIMS cable without PVC sheath carry out the following:

1. Strip back the copper sheath from the ends and make sure nothing is touching.

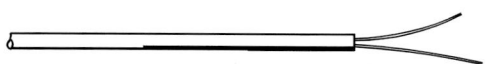

2. Connect the two separate cores to a bell and battery as shown.

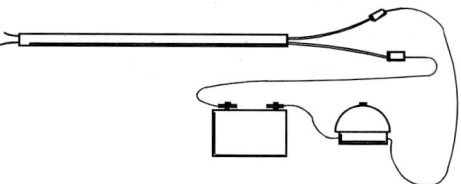

3. Heat up the cable in a blowtorch flame until it glows red hot. Remove the source of heat and cool by pouring water on the heated up cable.

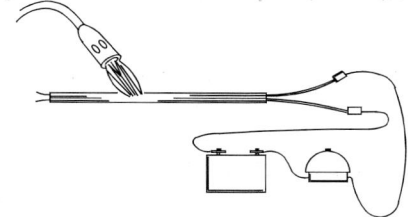

4. Now lay the cable on a solid metal surface and keep hitting it with a hammer until it is flat.

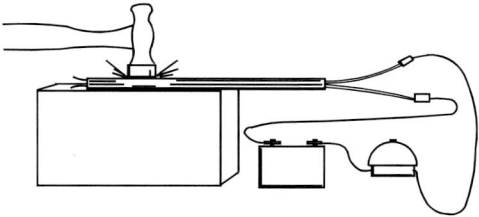

What condition was the cable in when the bell started to ring!

Self-assessed multi-choice questions.
Circle the correct answer in the grid below.

1. Conductors having a large c.s.a. are made up of several strands in order to
   (a) improve conductivity
   (b) reduce resistance
   (c) improve heat dissipation
   (d) give greater flexibility

2. Conductors having a very large number of fine strands are used for
   (a) overhead cables
   (b) underground cables
   (c) supplying large mobile machines such as cranes
   (d) fixed wiring in buildings

3. Aluminium cables of large cross-sectional area may have solid conductors.
   The reason for this may be
   (a) that aluminium is more flexible than copper
   (b) that this cable is to be laid underground and not subjected to movement
   (c) solid conductors have less resistance than stranded ones
   (d) they are stronger and less liable to breakage

4. Which of the following is not a standard cable size
   (a) 10 mm^2
   (b) 20 mm^2
   (c) 25 mm^2
   (d) 35 mm^2

5. A 100 A circuit is wired in 25 mm^2 copper cable. The same circuit if wired in aluminium cable would need a conductor size of at least
   (a) 16 mm^2
   (b) 25 mm^2
   (c) 35 mm^2
   (d) 50 mm^2

6. Which of the following cables is most likely to be installed in conduit or trunking?
   (a) mineral-insulated metal-sheathed
   (b) PVC steel wire-armoured
   (c) PVC-sheathed twin and multi-core
   (d) PVC singles

7. Which of the following would you consider most appropriate for a steel conduit installation?
   (a) a domestic dwelling
   (b) a farm building with livestock
   (c) a workshop
   (d) a car wash

8. Where conductor operating temperatures are likely to be too high for PVC steel wire-armoured cable a suitable alternative might be
   (a) XLPE
   (b) heavy duty PVC
   (c) PVC singles in conduit
   (d) PVC-sheathed non-armoured

9. Where PVC insulated or sheathed cable has been kept at a temperature below 0 °C it should be
   (a) immersed in hot water immediately
   (b) heated with a blowtorch
   (c) returned to the manufacturer
   (d) left in a higher temperature until safe to handle

10. The insulating medium in a MIMS cable is
    (a) mineral oxygen
    (b) magnesium oxide
    (c) magnesium
    (d) manganese

Answer grid

1	a	b	c	d	6	a	b	c	d
2	a	b	c	d	7	a	b	c	d
3	a	b	c	d	8	a	b	c	d
4	a	b	c	d	9	a	b	c	d
5	a	b	c	d	10	a	b	c	d

3

The Installation of Cables

You will need to have a current copy of IEE Guidance Note 1 and BS 7671 available for reference in order to complete the exercises within this chapter.

Take a few minutes to confirm that you have remembered the following important points:

The component elements of electrical cables include

The most commonly used conductor materials are

The resistance of conductors is determined by

On completion of this chapter you should be able to:

◆ calculate the minimum radius of cable bends allowable
◆ determine the maximum number of cables that can be installed in conduit without damage to their insulation
◆ calculate the current taken by equipment
◆ determine the method of cable insulation
◆ show the need to determine the ambient temperature
◆ take into account the grouping of cables
◆ calculate the cross-sectional area of a cable to meet given requirements
◆ complete the revision exercise at the beginning of the following chapter

Part 1

Supports, bends and space factors

Supports
Every cable needs to be installed in such a way that it is not subjected to undue stress or strain.

This means that the cable has to be adequately supported throughout its length (Figure 3.1).

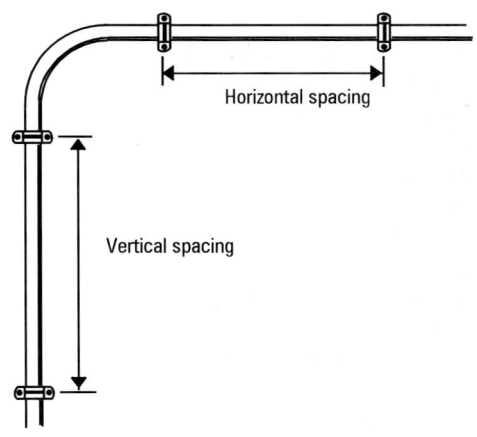

Figure 3.1

As a guide to good practice the IEE Guidance Note 1 gives maximum distances between clips.

Remember
The distances shown in IEE Guidance Note 1 are the maximum distances between supports and there is nothing wrong with using shorter intervals, especially where you think the cable might need additional support.

Where a cable is suspended overhead between two buildings a galvanized steel catenary wire is used to carry the weight of the cable (Figure 3.2). The cable can be fixed to the catenary wire using special cable hangers at the same spacings as for horizontal clips.

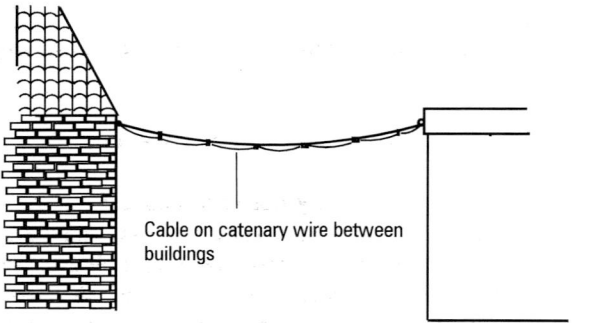

Cable on catenary wire between buildings

Figure 3.2

Vertical runs in conduit and trunking

Cables run in horizontal conduit or trunking do not need any support. Armoured cables lying on the bottom of ducts can be similarly considered.

Problems do arise, however, with vertical runs, because a long length of cable hanging down is going to put a considerable strain on itself where it bends over at the top.

For this reason, cables in conduit must have some form of intermediate support (Figures 3.3 and 3.4) where the vertical drop is more than 5 metres. The same applies to cables in vertical ducts and trunking.

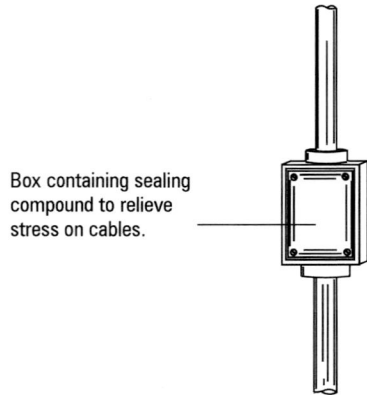

Box containing sealing compound to relieve stress on cables.

Figure 3.3   Vertical conduit

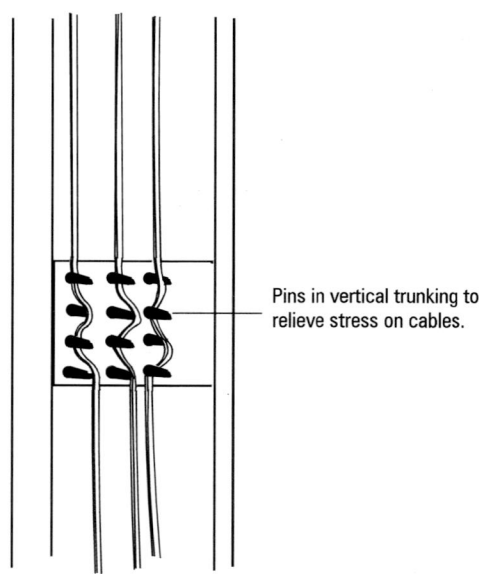

Pins in vertical trunking to relieve stress on cables.

Figure 3.4   Vertical trunking

Although the cables inside conduit and trunking are usually provided with ample support, these in themselves have to be fixed at certain maximum intervals. This will be covered in more detail later on in this book.

Bends

If a cable is bent too sharply it can suffer damage. This will affect the performance of a cable and could result in a dangerous situation.

A skilled electrician knows almost instinctively when a bend "looks right" and seldom has to check the fact. Those who have not yet acquired this skill will have to look up the table and satisfy themselves that their bends are not too tight.

The limiting factor is the minimum internal radius. This is half of the diameter of a circle which lies inside the bend, as shown in Figure 3.5.

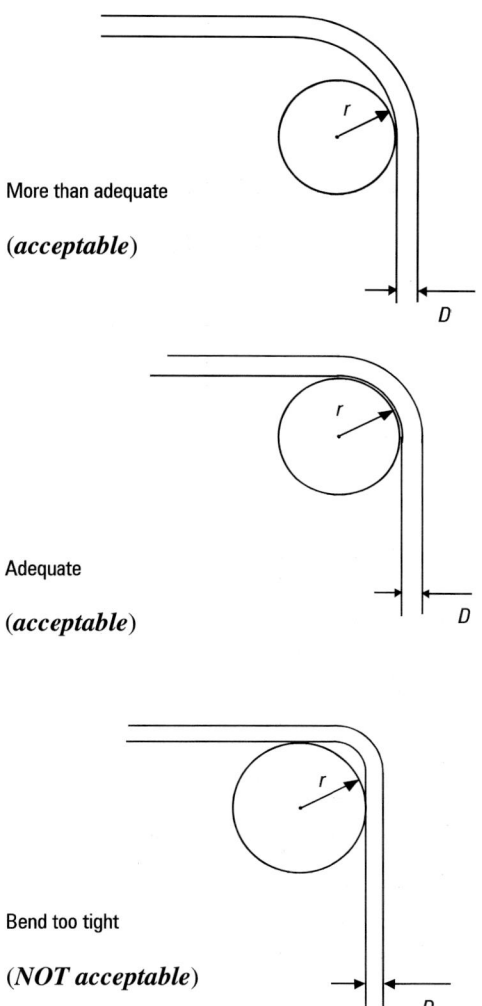

More than adequate

(*acceptable*)

Adequate

(*acceptable*)

Bend too tight

(*NOT acceptable*)

Figure 3.5

The minimum internal bending radius is found by multiplying the cable diameter by a factor. These factors can be found in IEE Guidance Note 1. For example, you will find that the multiplying factor for mineral-insulated cable of any overall diameter is 6. For PVC-insulated armoured or non-armoured cable any overall diameter it is 6 unless solid aluminium or shaped copper cores are used, when it is a factor of 8.

Try this

Take a piece of stiff card and draw two circles with a pair of compasses, one at 50 mm radius and the other at 62.5 mm. Cut out as shown and keep the card to check the bends on the two most popular sizes of conduit (20 mm and 25 mm).

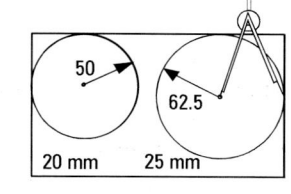

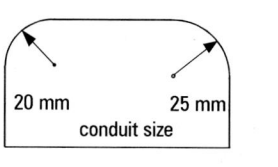

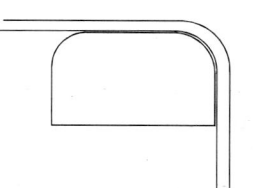

Example:

A mineral-insulated cable 10 mm in diameter has a multiplying factor of 6.

Minimum internal bending radius is $10 \times 6 = 60$ mm.

Bends in conduit are generally not less than 2.5 times the outside diameter of the tube, but you must always make sure that this is not less than the minimum bending radius of any cables to be drawn in after erection. Bends in trunking should always take into account the bending radii of the cables they contain.

A shaped piece of plastic, such as a pipe section, placed over the inside of a trunking bend will ensure that cables are not bent too sharply (Figure 3.6). This should be sized for the largest conductor to be installed.

This is especially appropriate for cables at the top of vertical drops (up to 5 m).

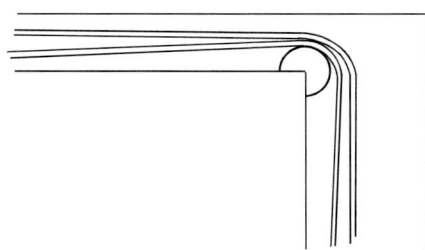

Figure 3.6

Space factor

Figure 3.7   *Do not try to put too many cables into an enclosure!*

You should never attempt to put too many cables into any enclosure because this could result in some of the insulation being damaged during installation (Figure 3.7). The factor system used to calculate how many cables can be put in an enclosure will be dealt with in Chapter 6, but as a general rule, the 45% space factor can apply.

This means that the cables should not occupy more than 45% of the available space (Figure 3.8).

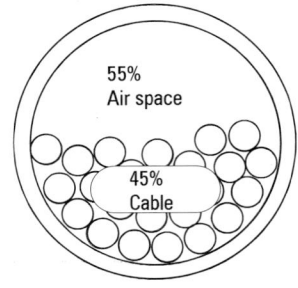

55%
Air space

45%
Cable

Figure 3.8

Example:

The inside diameter of a 20 mm conduit is found to be 17.5 mm.

Space available $\quad = \quad \dfrac{\pi d^2}{4} = 240 \text{ mm}^2$

45% of this space $\quad = \quad \dfrac{240 \times 45}{100}$

$\qquad\qquad\qquad = \quad 108 \text{ mm}^2$

Overall diameter of 1 cable including insulation
$\qquad\qquad\qquad = \quad 2.9 \text{ mm}$

Overall c.s.a. of 1 cable

$\qquad\qquad\qquad = \quad \dfrac{\pi d^2}{4} = 6.6 \text{ mm}^2$

Maximum number of cables

$\qquad\qquad\qquad = \quad \dfrac{108}{6.6}$

$\qquad\qquad\qquad = \quad 16$

Remember

Remember
Remember that to use the space factor the overall cross-sectional area of the cable including the insulation must be used.
You cannot use 1 mm^2, 2.5 mm^2 and so on because these refer to the conductor only.
The effective cross-sectional area of a non-circular cable is taken as that of a circle of diameter equal to the major axis of the cable.

Terminations

All terminations (Figures 3.9–3.13) should comply with certain conditions. These are:

- a termination should be electrically and mechanically sound
- the cable should be fully protected right up to the enclosure of the termination
- there should be no appreciable mechanical strain on the termination
- all terminations of conductors including live conductors should provide good continuity
- the insulation should be intact and undamaged right up to the point of the termination

There are a number of environmental considerations that must be given when deciding on a cable for a particular situation.

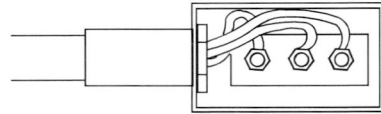

Figure 3.9   *Always allow enough "slack" in the termination*

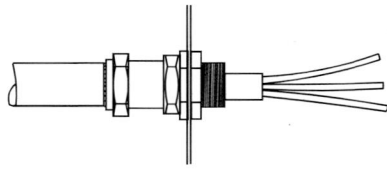

Figure 3.10   *SWA termination using proprietary weatherproof gland*

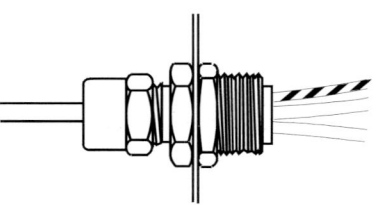

Figure 3.11   *MIMS termination using proprietary gland and seal*

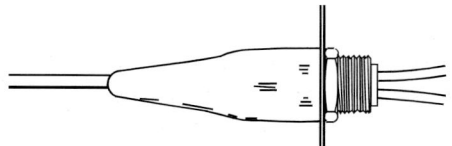

Figure 3.12   Use a PVC shroud to protect the gland

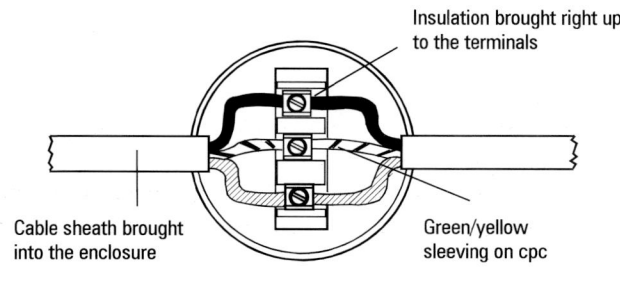

Insulation brought right up
to the terminals

Cable sheath brought
into the enclosure

Green/yellow
sleeving on cpc

Figure 3.13   Cable termination at joint box

Remember
**A mineral insulated cable must be sealed
to prevent moisture entering the cable.**

Points to remember ◄ – – – – – – – – – – – –

All cables must be supported so that they are not damaged. The spacing of the supports depends on the size of the cables and where they are fitted. The bending of cables also depends on the size and type of cable. A cable bent too tightly could cause the _____ to be damaged.

All terminations should comply with certain conditions.
* a termination should be _____ and mechanically sound
* the cable should be fully protected right up to the _____ of the termination
* there should be no appreciable _____ _____ on the termination
* where necessary the termination should provide good _____ continuity
* the _____ should be intact and undamaged right up to the point of the termination

It is important to use the correct type of termination for the cable.

Try this
Use a short length of 20 mm steel conduit and measure the inside diameter.

Inside diameter   _____   mm

Find a short length of 2.5 mm^2 PVC-insulated single cable and measure the outside diameter.

Outside diameter   _____   mm

Calculate the cross-sectional area of the inside of the conduit and the complete cable.

Area of conduit   _____   mm^2

Area of cable   _____   mm^2

Calculate how many 2.5 mm^2 cables can be installed in 20 mm conduit allowing for a 45% space factor.

Cut short lengths of 2.5 mm^2 cable to the number you have just calculated and insert them side by side in the conduit.

Self-assessment multi-choice questions.

Circle the correct answers in the grid below.

1. Using IEE Guidance Note 1, determine the recommended maximum distance between clips in a vertical run of PVC non-armoured cable 12 mm in diameter. Is it
   (a) 250 mm?
   (b) 300 mm?
   (c) 400 mm?
   (d) 450 mm?

2. Cables in conduit must have some form of intermediate support where the vertical drop is more than
   (a) 2 metres
   (b) 3 metres
   (c) 4 metres
   (d) 5 metres

3. Using the information given in this chapter the suggested minimum bending radius for a mineral insulated cable 14 mm in diameter is
   (a) 56 mm
   (b) 70 mm
   (c) 84 mm
   (d) 98 mm

4. The number of 4.3 mm diameter cables which can be drawn into conduit with an inside diameter of 22 mm without exceeding the 45% space factor is
   (a) 11
   (b) 13
   (c) 15
   (d) 17

5. The minimum bending radius for a 32 mm conduit would be
   (a) 50 mm
   (b) 62.5 mm
   (c) 80 mm
   (d) 100 mm

Answer grid

1	a	b	c	d
2	a	b	c	d
3	a	b	c	d
4	a	b	c	d
5	a	b	c	d

Remember

Before making your choice of wiring system, consider all the environmental factors, such as:

Temperature
  Is it going to be very hot or very cold?

Humidity or moisture
  Is it likely to get damp due to water or condensation?

Are there any other possible hazards?
  Oils, fats, solvents, corrosive substances.
  Fire or explosion risks and so on.

Mechanical damage
  Movement of machinery or materials, possibility of vandalism, and so on

Then look at the proposed method of installation:
  Concealed in the building's structure
  Clipped directly to the surface
  Enclosed in conduit or trunking
  Clipped to cable tray
  Laid in ducts
  Suspended overhead

But don't forget the details, such as:
  Access to terminations
  Bends
  Number of cables in an enclosure
  Availability of fixings for proper spacing

Did you know that perforated cable tray could solve your support problems where direct fixings are difficult to make?

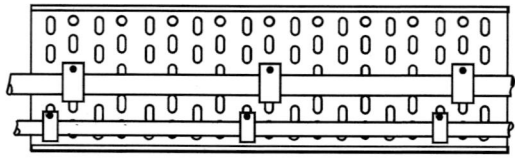

Part 2

Cable selection

Back in Chapter 2 you completed a table of current rating against cross-sectional area. On the face of it there doesn't seem to be much of a problem.

Selecting a cable means
- working out the load current
- looking up a cable
- picking out a cable which is big enough for the current

That is still true but, as you may have guessed, there's a lot more to it than that.

Figure 3.14

Let's deal with the first part first.

Working out the load current

This is called the "circuit design current" and is given the symbol Ib.

For a single load, you might find this information on a maker's plate (Figure 3.15).

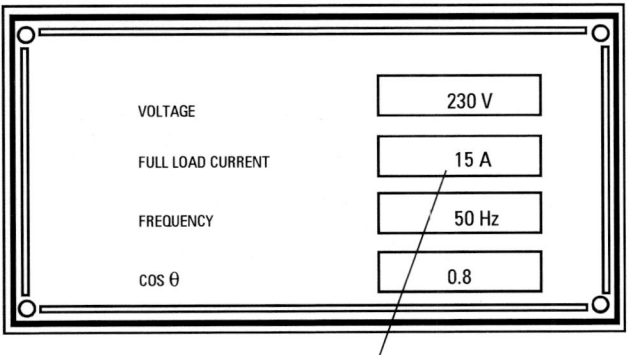

VOLTAGE	230 V
FULL LOAD CURRENT	15 A
FREQUENCY	50 Hz
cos θ	0.8

This is what you are looking for.

Figure 3.15

For a single load in kilowatts it's not too difficult.

Simply

$$\frac{\text{Divide the load in kilowatts} \times 1000}{\text{By the mains voltage} \times \text{the power factor}}$$

usually written

$$I = \frac{P}{U_0 \cos \theta}$$

Now let's explain the power factor

Power factor
(Symbol: $\cos \theta$).

In some a.c. circuits the current and voltage get out of step with each other and don't pull together as they should.

This reduces the amount of power you are going to get for that current and voltage, or alternatively you've got to draw more current from the supply in order to get the power you need. Either way it affects the calculated value of load current.

Here is a picture example (Figures 3.16 and 3.17).

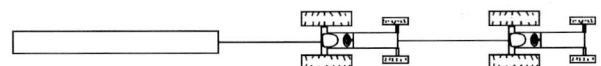

Figure 3.16   *Two tractors hauling a log put maximum effort into the tow rope when pulling in tandem.*

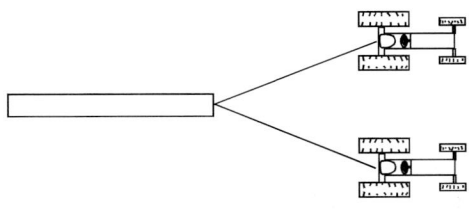

Figure 3.17   *Two tractors still exert force on the log, but with some reduction in effort because neither is exerting a straight pull.*

Example

What is the current drawn by a 6 kW load with a power factor of 0.6 connected to a 200 V supply?

$$\text{Load current} = \frac{\text{kilowatts} \times 1000}{\text{voltage} \times \cos\theta}$$

$$I = \frac{P}{V\cos\theta}$$

$$= \frac{6000}{200 \times 0.6}$$

$$= \frac{6000}{120}$$

$$= 50 \text{ amperes}$$

If a load has no stated power factor, this may be because there is no phase difference between the current and the voltage. This happens with tungsten lighting and heating loads, and in such cases the power factor ($\cos\theta$) is 1.

The power factor is always 1 for all direct current (d.c.) loads.

Try this

1. A 230 V, 4.6 kW load has a power factor of 0.8. What is its load current?

2. A tungsten lighting circuit consists of ten 230 V, 100 W luminaires, each with a power factor of 1 (unity). What is the circuit design current?

Reference method

The cable you eventually choose will have to be selected from a table and unless you select it from the correct table of current ratings then your choice is likely to be wrong. Tables are drawn up for lots of different types of cable but even within that table there are several different columns according to the method of installation. Each method is given a reference number. There are twenty of them but we'll deal with only three examples (Figure 3.18). BS 7671 Table 4A is where you will find the complete list.

Remember

"R"

reminds you to look up the reference method.

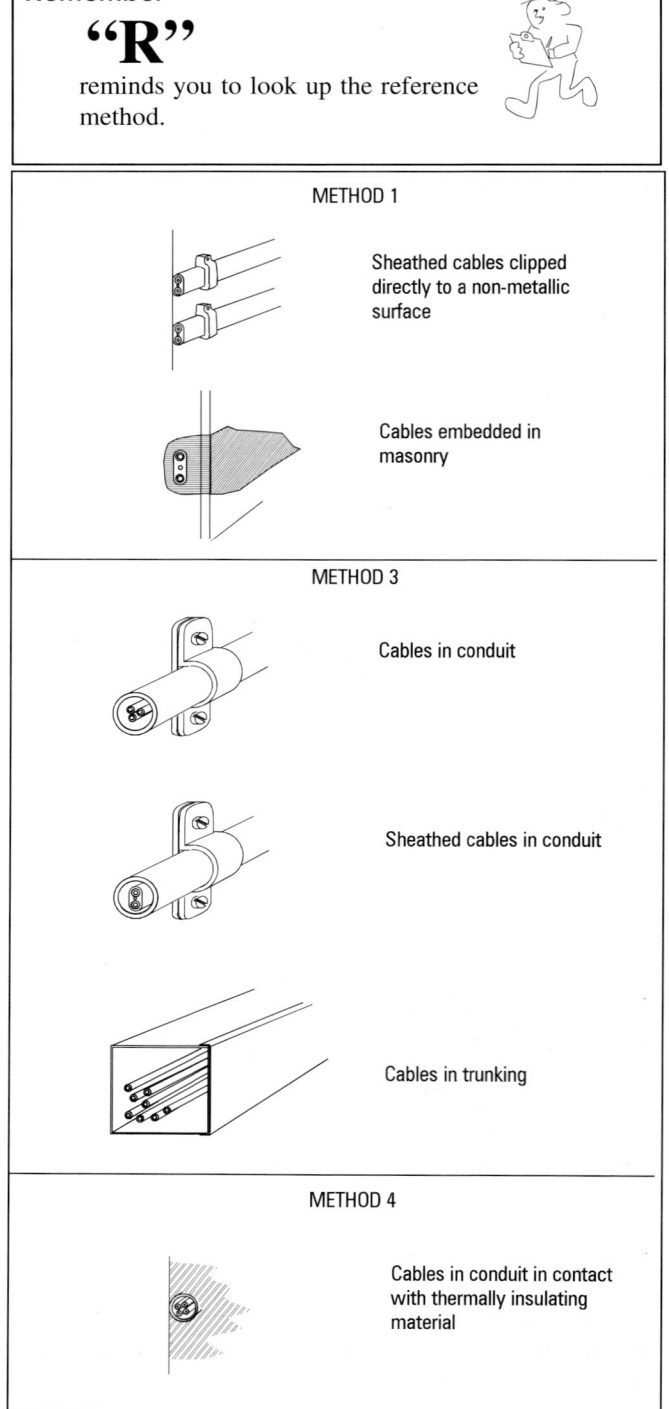

METHOD 1

Sheathed cables clipped directly to a non-metallic surface

Cables embedded in masonry

METHOD 3

Cables in conduit

Sheathed cables in conduit

Cables in trunking

METHOD 4

Cables in conduit in contact with thermally insulating material

Figure 3.18

Type of overcurrent protection

Some kinds of fuses will "operate" (that's what we say when we mean blow or trip) when the circuit current exceeds their current rating by a relatively small amount.

This also applies to miniature circuit breakers because they have a design characteristic which makes them do this.

Fuses and circuit breakers in this category are usually referred to by their British Standard Specification numbers.

You will probably recognise them by sight (Figures 3.19–21).

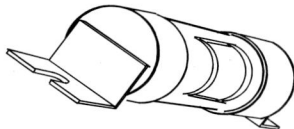

Figure 3.19   *BS 88 Part 2 and Part 6*

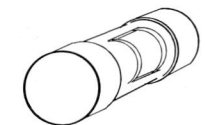

Figure 3.20   *BS 1361 and 1362*

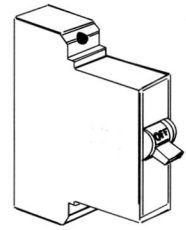

Figure 3.21   *Miniature circuit breaker BS EN 60898*

However, problems arise when you've got one of the fuses in Figure 3.22 protecting the cable.

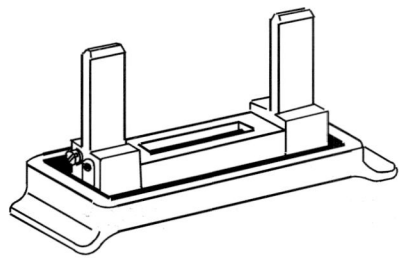

Figure 3.22   *BS 3036 semi-enclosed fuse*

This is a BS 3036 **semi-enclosed** rewireable fuse.

This one doesn't operate until the load current is about TWICE the rating of the fuse wire element. To compensate for this we need to apply a factor for this type of device.

There's no need to look up a table because the factor for this fuse is always 0.725.

Thermal insulation

Where cables are to be run in a building void which is going to contain thermal insulation, the cable should be fixed in such a way that it is not completely surrounded.

This means that Reference Method 4 applies.

If the only route available takes the cable through the insulation (Figure 3.23), then why not place the cable in conduit or trunking and apply Method 4 as before?

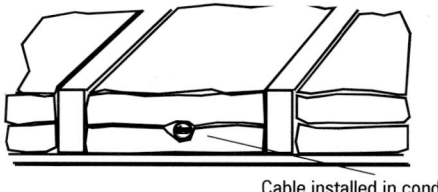

Cable installed in conduit

Figure 3.23   *Cable installed in conduit in thermal insulation*

If, after giving it a lot of thought, you have decided that the cable must pass through the insulation and become totally enclosed then you must apply the factor given in BS 7671 Regulation 523-04-01 using Reference Method 1. But remember that PVC cables directly in contact with some thermal insulation material will not only be affected by the heat produced, but also from a chemical reaction which may damage the cable.

That is, if the total length of cable enclosed in insulation is 0.5 m or more a factor of 0.5 must be applied to the Method 1 ratings.

For lengths up to 0.5 m see Table 52A in BS 7671.

Ambient temperature

When a cable carries current it will begin to get warm. This is not unusual and in most cases creates no problem.

If the temperature of the surroundings (Figure 3.24) is already high, then this additional heat might push the cable towards its temperature limit. So what do we do to prevent a cable from overheating in these circumstances?

Use a bigger cable. How big? Depends on the temperature.

How do we find out? Look up another table, this time BS 7671, Table 4C1. Where the circuit is protected by a semi-enclosed (rewireable) fuse, Table 4C2 applies.

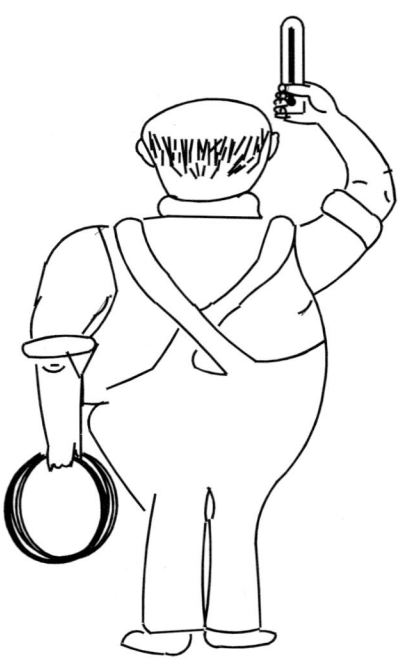

Figure 3.24   Checking the ambient temperature

Remember

If the cable is run in an area of high ambient temperature then we must

1. Find out the temperature.

2. Look up the factor.
(Note that for a temperature of 30 °C the factor is 1.0 which does not affect the rating of the cable.)

3. Write it down and call it

"A"

Grouping

A twin cable, or a pair of singles will generate a bit of heat but this need not be too much of a problem if the heat can escape into the surrounding air.

What frequently happens though, is that a single enclosure, possibly a conduit or trunking, has several circuits all bunched together, all similarly loaded and all trying to get rid of their own heat (Figure 3.25).

The effect is obvious. The combined heating effect is going to raise the temperature of the group.

This means that unless you have taken steps to prevent it, all the cables are going to go over their temperature limit.

So what are we going to do?

Look up a table – that's what.

This time it's BS 7671 Table 4B1.

Figure 3.25   How many in the group?

Remember
Grouping

1. Find out how many are in the group.

2. Look up the factor.

3. Write it down and call it

"G"

Try this

You will need to refer to a current copy of BS 7671 to be able to enter the appropriate factors in the following tables. Note the number of the table in BS 7671 from where you obtained the information.

1. (a) Factors for general-purpose PVC-insulated cable involving ambient temperature where the circuit is protected by a semi-enclosed (rewireable) fuse.

25	30	35	40	45	50	55	60	65	Temp °C
									Factor

(b) Factors for general-purpose PVC-insulated cable involving ambient temperature other than in (a).

25	30	35	40	45	50	55	60	65	Temp °C
									Factor

2. Grouping factors

Reference method Enclosed (3 or 4)	number of circuits of multi-core cables						
	2	3	4	5	6	7	8
Factor							

Figure 3.26

Let's get it all together

First

check up on the load current. Once you've decided what it's going to be, call it

*I*b

and

select a fuse or circuit breaker of the type you're going to use, such that its rating is either equal to, or just larger, than Ib, and call it

*I*n

then

look at the way in which the cable is going to be installed and decide on the appropriate reference method

now

if the fuse is a semi-enclosed rewireable BS 3036 type apply the factor 0.725

S

then

see if the cable is going to be totally enclosed in thermal insulation and write down the factor

T

then

find out the ambient temperature of the location. Look up BS 7671 Table 4C1 or 4C2, whichever, applies and note the factor

A

and finally

see if your cable is going to form part of a group. If it does, look up the grouping factor

G

Let's see how it works

Say, for example, the load current is 12 amperes.

$$Ib = 12 \text{ A}$$

The circuit is protected by a BS 1361 fuse so we'll need a rating of 15 amperes

$$In = 15 \text{ A}$$

Table 3.1

BS 1361 fuses available				
Rating	5 A	15 A	20 A	30 A	45 A

The cable is to be installed in steel conduit. Reference Method 3.

Reference Method 3

Semi-enclosed fuse in *not* used so

S is 1

No thermal insulation problem so

T is 1

The temperature of the room is 35 °C.
Look up BS 7671, Table 4C1, the factor for general-purpose PVC insulated cable.

A is 0.94

There is a total of three similar circuits in the conduit.
Look up BS 7671, Table 4B1.

G is 0.7

Now:

we divide the fuse rating (*In*) by any factors which may apply, and this will give us an idea how big the cable's got to be.

The tabulated current carrying capacity must not be less than ($\geq$)

$$\frac{\text{fuse rating}}{S \times T \times A \times G}$$

$$\frac{15}{1 \times 1 \times 0.94 \times 0.7}$$

22.8 A

Now we look for a cable in the table of PVC Singles under Reference Method 3 which is rated at not less than 22.8 A. (Look this up in BS 7671 if you have available a copy for reference).

From this we can see that the cable you need is 2.5 mm², which has a tabulated current rating of 24 A (Table 4D1A, column 4).

In BS 7671 the symbol C_i is used for the thermal insulation factor, C_a for the temperature factor and C_g for the grouping factor.

Try this

You will need to refer to a current copy of BS 7671 in order to complete this exercise. Select a cable from BS 7671, Reference Method 3, which can supply a load current of 28 amperes. It is to be wired in PVC singles in steel conduit as part of a group of 4 circuits in an ambient temperature of 40 °C and is protected by a BS 3036 semi-enclosed fuse.

BS 3036 fuses available				
Rating	5 A	15 A	20 A	30 A	45 A

Try this

Try these power factor questions.

1. What is the current drawn by a 9.2 kW load with a power factor of 0.8 connected to a 230 volt supply?

2. If the current drawn by a 3.2 kW load connected to a 200 volt supply is 20 A what is the power factor?

3. What is the current drawn by a 9 kW load with a power factor of 0.75 connected to a 230 volt supply?

4. If the current drawn by a 6 kW load connected to a 200 V supply is 30A what is the power factor? Why could this be?

The tables below, which have been mentioned in this chapter, can be found in either BS 7671 (Requirements for Electrical Installations), manufacturers' catalogues or in the Guidance Notes published by the IEE. Make a note below where you can find complete tables on the following:

Spacing of supports for cables

Spacing of supports for conduits

Minimum internal bending radius

Correction factors for ambient temperature

Current-carrying capacity for copper conductors

Self-assessment multi-choice questions.

Circle the correct answers in the grid below.

1. The Reference Method used for cables installed in conduit fixed to the surface is
   (a) Method 1
   (b) Method 2
   (c) Method 3
   (d) Method 4

2. The correction factor for ambient temperature does not affect the rating of cables at
   (a) 0 °C
   (b) 20 °C
   (c) 30 °C
   (d) 40 °C

3. A correction factor of 0.725 is applied when the overcurrent protection is provided by a
   (a) BS 88 fuse
   (b) BS 1361 fuse
   (c) BS EN 60898 m.c.b.
   (d) BS 3036 fuse

4. The current drawn by a 230 V 2.3 kW load with a power factor of 0.8 is
   (a) 12.5 A
   (b) 10 A
   (c) 8 A
   (d) 3 A

5. The load current can sometimes be described as the "circuit design current" and is given the symbol
   (a) I_d
   (b) I_t
   (c) I_c
   (d) I_b

Answer grid

1	a	b	c	d
2	a	b	c	d
3	a	b	c	d
4	a	b	c	d
5	a	b	c	d

Part 3

Variations in conditions

The conditions which bring about the need for a correction factor may exist throughout the whole length of a cable run, in which case all factors apply.

But take for example a cable with general-purpose PVC insulation which is protected by a BS 3036 fuse (Figure 3.27).

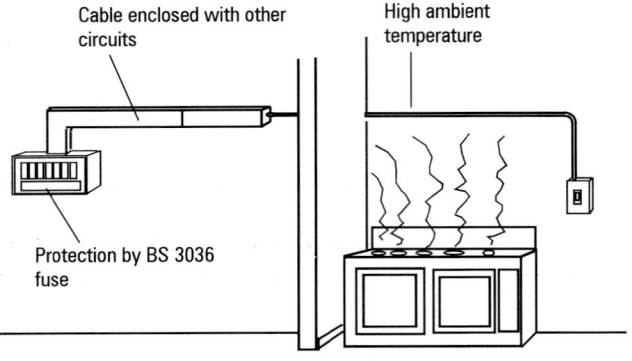

Figure 3.27

Example

S of 0.725 applies throughout.

The cable is not completely enclosed in thermal insulation at any stage.

$T = 1$

On the next part of the run it is in an ambient temperature of 45 °C but run singly.

$A = 0.91$ (4C2)

For part of the run it is in a group of 4 but no other factors apply.

$G = 0.65$ (4B1)

In these circumstances the "worst case" prevails.

First part
$S \times T \times A \times G = 0.725 \times 1 \times 1 \times 0.65 = 0.47$

Second part
$S \times T \times A \times G = 0.725 \times 1 \times 0.91 \times 1 = 0.66$

The worst case is the first part with a combined correction factor of 0.47.

The cable selected for the first part will therefore satisfy the worst conditions to be encountered throughout the run.

Let's try another.

Example
PVC singles in conduit, Reference Method 3.

Load current is 8 A.

$I_b = 8$ A

This needs a 10 A fuse.

$I_n = 10$ A

But in this case the fuse is a rewireable one (BS 3036).

$S = 0.725$

No contact with thermal insulation

$T = 1$

The ambient temperature is 30 °C.

$A = 1$

and there is one other circuit in the conduit (making a group of 2)

$G = 0.8$

$$I_t \geq \frac{I_n}{S \times T \times A \times G} \geq \frac{10}{0.725 \times 1 \times 1 \times 0.8}$$

Tabulated current ≥ 17.24 A

Cable selected: 1.5 mm^2 with a capacity of 17.5 A.

Figure 3.28

Voltage drop

In this circuit (Figure 3.29), a current of 5 A flows through a twin cable with a total resistance of 2 Ω.

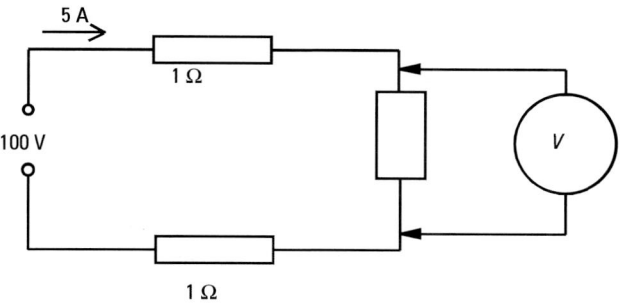

Figure 3.29

If you remember your Ohm's Law then you'll also remember that

$$V = IR$$

in this case $V = 5 \times 2 = 10$ volts

The supply voltage is 100 V.

There is a 10 V drop in the cable.

Therefore the voltage at the load is 90 volts.

BS 7671 requires that the total voltage drop between the supply position and the fixed current using equipment should not exceed 4% of the nominal supply voltage.

In practical terms the maximum permissible voltage drop in a 230 volt installation is

$$\frac{230 \times 4}{100} = 9.2 \text{ volts}$$

Where this represents the sum of the voltage drops in more than one cable the actual voltage drop in each cable becomes a matter for the designer.

The way in which we normally calculate voltage drop is to use the same tables as we used for the current carrying capacity of cables (BS 7671).

The mV/A/m quoted are millivolts per ampere per metre run.

By using these tables you can calculate the actual voltage drop in any given cable provided that you know
- the length of run
- the circuit design current I_b
- the voltage drop in mV per ampere per metre.

Cable selection

If a cable has to satisfy the requirements of BS 7671 for current-carrying capacity and voltage drop then you must check to ensure that both requirements are met.

In the first example shown on the facing page the voltage drop is 5.76 volts.

In the second example the voltage works out at 7.26.

When it comes to cable selection you haven't really got much choice.
- You can't change the load current.
- You can't change the length of the run.
- But you can choose a bigger cable if you need to.

So select a cable for its current carrying-capacity then check to see that the voltage drop is within the required figure.

But remember that the voltage drop is taken from the incoming supply terminals to the farthest point in the circuit and the voltage drop you are looking at may only be a part of the total 9.2 volts permissible.

In Figure 3.30 you see that each successive circuit has a maximum voltage drop of 3.06 volts (or 3.06 × 1000 millivolts).

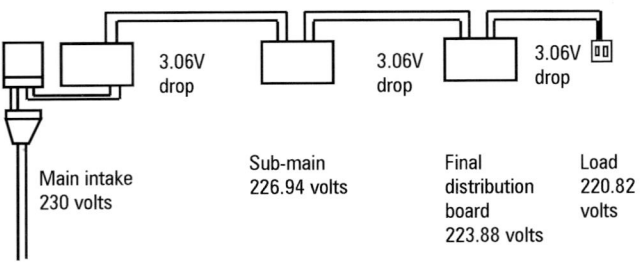

Figure 3.30

So each cable would have to be selected accordingly, i.e.

$$\text{mV/A/m} \leq \frac{3060}{I_b \times \text{length of run}}$$

Example

A 20 m run of 2.5 mm^2 twin cable has a load current of 16 A. What is the voltage drop? From Table 4D2B in BS 7671 the mV drop for 2.5 mm^2 is **18 mV/A/m**.

Voltage drop =

$$\frac{\text{load current} \times \text{length of run in metres} \times \text{mV drop}}{1000}$$

$$= \frac{16 \times 20 \times 18}{1000}$$

$$= 5.76 \text{ volts}$$

Example

length of run	=	22 metres
load current	=	30 A
c.s.a of cable	=	4 mm^2

(from the table 4 mm^2 has a mV drop of 11)

Voltage drop	=	7.26 volts

You must always divide by 1000 to convert your answer from millivolts to volts.

Selection by voltage drop

It is quite easy to make a cable selection using the voltage drop.

Maximum permissible mV drop =

$$\frac{\text{maximum permissible volts drop} \times 1000}{\text{load current} \times \text{length of run}}$$

Taking an earlier example:

length	= 20 m
load current	= 16 A
maximum permissible voltage drop	= 3.06 V
	(3060 mV)

$$\text{maximum mV drop} = \frac{3060}{20 \times 16} = 10$$

$$= 9.6 \text{ mV/A/m}$$

Now, using the appropriate table in BS 7671, look down the column until you find the c.s.a of cable which has a mV drop figure less than the one you've just calculated.

The cable chosen should be 6 mm^2.

This may not be an ideal choice and it may then be left to the designer to choose another voltage drop within the total 9.2 volt limit.

Now, taking the second example, the one which gave a voltage drop with 7.26 volts:

$$\text{maximum mV drop} = \frac{3060}{30 \times 22}$$

$$= 4.6 \text{ mV/A/m}$$

The cable chosen should be 10 mm^2 (4.4 mV/A/m).

Remember

The cable must satisfy the requirements of BS 7671 in respect of current carrying capacity *and* voltage drop.

Points to remember ◄ – – – – – – – – – – – – – –

In order to calculate the size (the cross-sectional area) of a cable required for a particular situation a number of factors have to be considered.

The load current must be determined (in some a.c. circuits this may involve the use of the _____ factor).

Now before you can look up the size of cable in the relevant table, other tables may have to be consulted in regard to
the method of installation
(known as the **R**eference Method)

whether the cable is protected by a
BS 3036 _____ (S)   **S**

If the cable is totally enclosed in thermally
_____ material a further correction factor
is applied (T)   **T**

the temperature of its surroundings (the **A**mbient
temperature)   **A**

the grouping of cables
(the **G**rouping Factor)   **G**

Where varying conditions are present in a cable run then the cable selected must satisfy the "_____
_____" encountered.

Cables can also be selected using the voltage drop but they must satisfy BS 7671 in respect of the current-carrying capacity.

Figure 3.31

Self-assessment multi-choice questions.

Circle the correct answers in the grid below.

1. If three twin cables are to be installed in the same trunking the appropriate correction factor will be
    (a) 0.725
    (b) 0.7
    (c) 0.61
    (d) 0.5

2. If cables are to be totally enclosed in thermal insulation for a distance greater than 0.5 m a correction factor is applied of
    (a) 0.3
    (b) 0.4
    (c) 0.5
    (d) 0.6

3. A PVC-insulated cable is required to supply a load of 10 A in an ambient temperature of 40 °C. No other factors apply and the fuse is 15A BS 1361. Reference Method 3. Using BS 7671 your chosen cable is
    (a) 1.0 mm^2
    (b) 1.5 mm^2
    (c) 2.5 mm^2
    (d) 4.0 mm^2

4. Using BS 7671 the voltage drop in 25 metres of 2.5 mm^2 twin cable at a load current of 15 A is
    (a) 0.148 volts
    (b) 6.75 volts
    (c) 14.8 volts
    (d) 6750 volts

5. So that the voltage drop in Question 3 does not exceed 5 volts over a length of 20 m the cable used must be
    (a) 1.0 mm^2
    (b) 1.5 mm^2
    (c) 2.5 mm^2
    (d) 4.0 mm^2

Answer grid

1	a	b	c	d
2	a	b	c	d
3	a	b	c	d
4	a	b	c	d
5	a	b	c	d

4

Enclosed Cable Systems

In the last two chapters we looked at cables. Note down the answers to the following questions before you continue.

What is the function of cable sheathing?

What external influences need to be taken into consideration when choosing a cable for a particular location?

What type of support (for example clips, conduit and so on) would you recommend for cables in vertical and horizontal runs?

Figure 4.1

The main reason for using any form of enclosed cable system is that it offers a high level of mechanical protection, yet at the same time allows the system to be easily altered or rewired. As the enclosure provides the mechanical protection for the cables it is not necessary to provide mechanical protection on the cable itself. Therefore the conductors are provided with electrical insulation only and are known as *single insulated cables*. It follows that composite cables, such as 2.5 mm^2 twin and cpc, are not used in enclosed systems as the combined cable is held together by the mechanical protection (the cable sheath) (Figure 4.2).

Insulated and sheathed single conductor

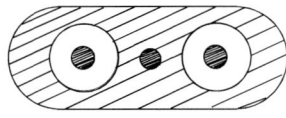

Insulated and sheathed composite cable

Single insulated

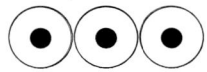

Equivalent single insulated

Figure 4.2

As we can see, the thickness of the electrical insulation is the same for both types of cable. If care is not used when installing single insulated cables any damage will be to the electrical insulation, so both the insulation resistance and voltage rating will be reduced as both of these are dependent on the thickness of the insulation.

As the enclosure provides the mechanical protection it is common for a large number of circuits to share a single enclosure. It is also common practice to use more than one type of enclosure in an enclosed system.

Common types of enclosure in general use

Conduit

Conduit (Figure 4.3) is basically a pipe into which the single insulated conductors are installed. These conduits are usually either steel or plastic. Although other materials such as aluminium are available, these are now seldom used except for some special installations.

Steel conduit is supplied in standard lengths of 3.75 metres and with outside diameters of 16 mm, 20 mm, 25 mm and 32 mm. These lengths are cut and joined as required, using a variety of connecting boxes to form a complete system of tubing to each point on the system, and into this the required cables are installed.

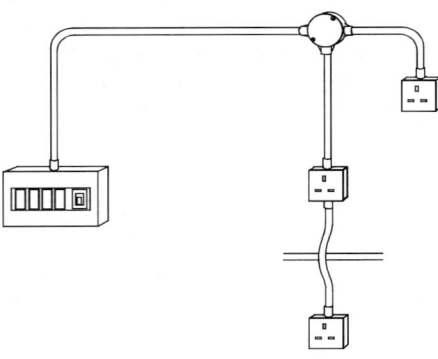

Figure 4.3

Trunking

Trunking is really a box section with a detachable lid. It is supplied in standard lengths of 3 metres and with dimensions between 38 mm × 33 mm and 225 mm × 100 mm. The trunking is usually connected together to form a continuous channel into which the cables are laid, and the lid is then fitted to complete the enclosure. It is common practice to use trunking as the main run for large numbers of cables and to take off from the trunking in conduit to supply individual items of equipment (Figure 4.4).

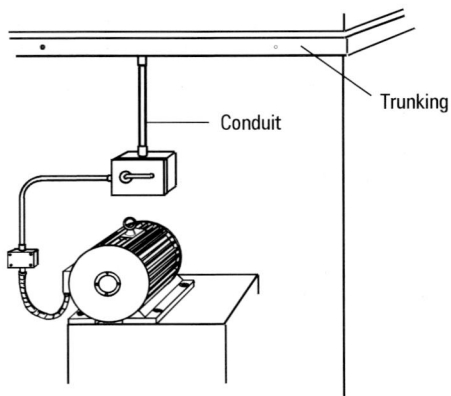

Figure 4.4

Use of conduit and trunking

The most likely place to find any enclosed system is in an industrial location, such as a factory, where mechanical protection is important and where a large number of circuits may be run sharing a common route.

If a large number of circuits share this common route and it is accessible then trunking may be used to form the enclosure. Take-off positions from this main run may be in smaller section trunking to smaller areas, conduit to individual machines and so on.

Special trunking may be used to allow luminaires (light fittings) to be suspended directly from the trunking in which the lighting cables are installed; this is known as lighting trunking (Figure 4.5).

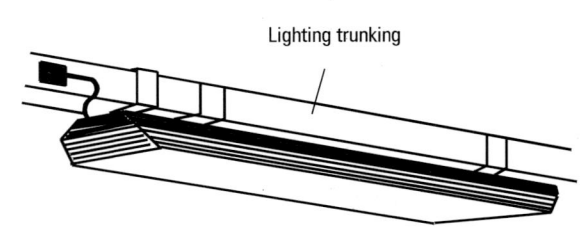

Figure 4.5

Conduit systems may also be used in hospitals, shops or blocks of flats when the conduits are chased into the walls and installed within the "cast *in situ*" concrete. They may be installed above ceilings and under concrete floors before they are laid. In this way the conduit system is completely hidden from view and is known as a "flush" conduit system. Steel conduit encased in cement or plaster, unless protected, will corrode, and as it is possible to use steel enclosures as a circuit protective conductor (cpc) steps must be taken to prevent this corrosion. Alternatively, plastic conduit could be used, which will not corrode, but a separate circuit protective conductor must be installed for each circuit. Plastic conduit has become widely used for such installations as it is both unaffected by the chemicals in the cement and plaster and it is very quick to install.

Before we look at these systems in detail there are some general points which we should consider which apply to all enclosed systems.

Storage of materials

As the enclosures come in fairly long lengths these may prove difficult to store, and in the case of steel, heavy and cumbersome to move. It is important, however, that the enclosures are supported correctly, as they may be bent or damaged easily. It is also vital to keep enclosures in an environment where they will be relatively unaffected. If, for example, steel conduit is stored out of doors, in damp situations or laid on concrete floors, then corrosion will set in. If plastic conduit is stored horizontally it will require solid support in order to prevent it bending.

PVC conduit is also likely to distort with changes in temperature and care must be taken to ensure that such material is stored at an appropriate temperature with adequate support to prevent this happening. Figures 4.6 and 4.7 show some suggested methods of storing conduit and trunking.

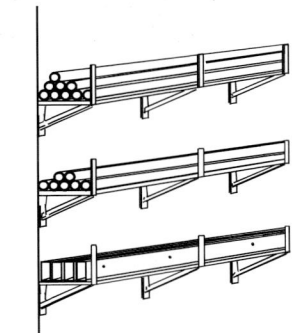

Figure 4.6   *Conduit and trunking stored horizontally*

Figure 4.7   *Conduit and trunking stored vertically.*

Cable drums should be stored stacked flat or on rods and clear of the floor. Care must be taken to ensure that no damage occurs to the cables or the drums whilst they are stored. They must also be kept dry as soggy damp cardboard drums will fall apart making the installation of cables difficult. Figure 4.8 shows a method of storing cables, conduit and trunking on site. Large wooden drums of cable will be stored on the floor, with wedges placed under the rim of the drums to stop them rolling.

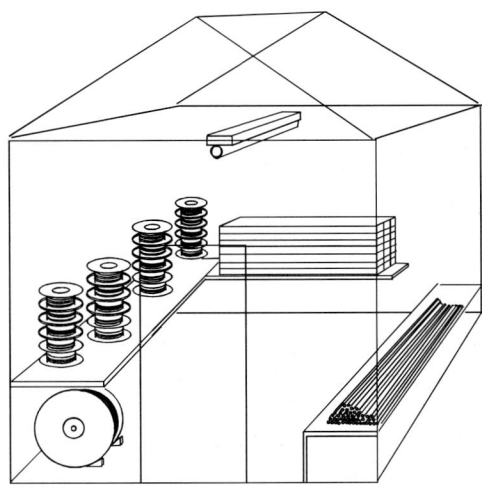

Figure 4.8   *Site storage hut*

Installing materials

A steel enclosure creates some problems for us with respect to both installation and operation. During assembly, conduit is cut and threaded and trunking cut and joined. If any rough edges are left, then as cables are drawn in these edges act like razors and remove slivers of insulation from the cables. To avoid this it is essential that all the cut edges are cleaned so that no burrs are left. This is done by filing for trunking and with a deburring tool for conduit, after cutting and before assembly (Figures 4.9 and 4.10).

The inside of the conduit
cleaned off with a reamer

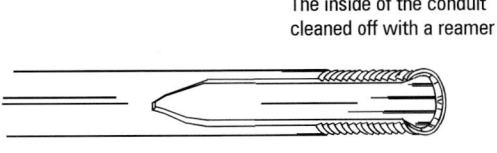

Figure 4.9   *A section across the end of a piece of conduit*

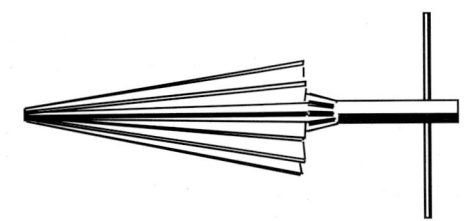

Figure 4.10   *Conduit reamer*

Once the conductors are installed we have a couple of other problems.

Firstly, if we pass an a.c. current through a conductor an alternating magnetic field is produced around the conductor. If the current flowing through all the conductors is in the same direction then the magnetic flux produced will combine to produce a large circulating magnetic flux. Steel is a ferrous material and can therefore be magnetised. This results in the flux produced circulating in the conduit or trunking. This in turn will induce a current in the steel enclosure and can cause a rise in temperature in the enclosure, as shown in Figure 4.11. Both of these effects are undesirable. To prevent them occurring it is important that for single phase circuits both phase and neutral conductors are run in the same enclosure and that all three-phase conductors, and any neutral conductors, for three-phase circuits, are contained in the same enclosure. This has the effect of ensuring that the current to and from the load is passed through the enclosure. As the currents are flowing in opposite directions, the magnetic flux produced will also be in opposite directions, so they will tend to cancel one another out. No flux will circulate in the conduit, so no current flow or temperature rise will occur.

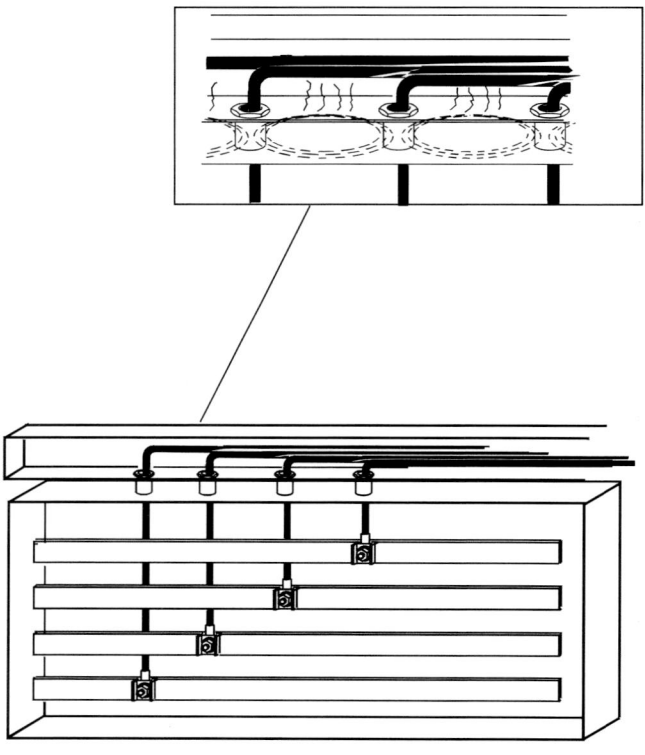

Figure 4.11   Undesirable electromagnetic effects

If circuit conductors are run through separate couplers, as shown in Figure 4.11, heat is produced in the steel. To eliminate this, all of these conductors should pass through a single coupling (Figure 4.12).

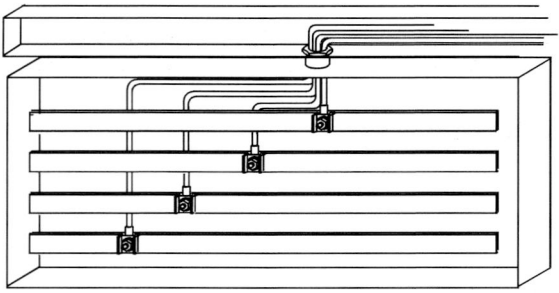

Figure 4.12   The conductors passing through a single coupling

The second problem is also connected with the magnetic effect of a steel enclosure. If the steel is used as a conductor, then when an a.c. current flows through it a magnetic field is set up which tends to oppose the flow of current through it. This is known as the inductive effect. As a result the overall impedance to a.c. current is higher than the resistance to d.c. current.

We may need to use a high-current, low-impedance ohmmeter to check the impedance of a steel cpc. This test is carried out at 50 V with a current approaching 1.5 times the design current of the circuit, subject to a maximum of 25 A.

It is, of course, vital that the impedance of the steel enclosure is sufficiently low throughout the system if it is to be used as the protective conductor for the circuits.

> ### Try this
>
> Take a look around your workplace and see if you can locate any enclosed systems.
>
> Note how many of the types we have mentioned are installed.

Points to remember ◄ – – – – – – – – – – – – – –

The enclosure forms the mechanical protection for the cables which have electrical insulation only around the conductor.

Conduit is usually made of _____

or _____

Trunking is connected together by

The most likely place to find any enclosed system is in an _____ location, for example a _____

Other likely locations for conduit systems are

Steel enclosures may be used as the circuit _____ _____ for the circuits installed.

Enclosures generally fall into two categories:

1.

2.

What happens if steel conduit is stored in a damp situation?

Try this

Write in your own words the effect that magnetic flux circulating in conduit or trunking has and also what can be done to overcome it.

47

Self-assessment multi-choice questions.

Circle the correct answers in the grid below.

1. Conduit is supplied in standard lengths of
   (a) 2 m
   (b) 2.75 m
   (c) 2.5 m
   (d) 3.75 m
2. Which of the following is not a standard diameter for steel conduit?
   (a) 15 mm
   (b) 20 mm
   (c) 25 mm
   (d) 32 mm
3. Trunking is supplied in standard lengths of
   (a) 2.25 m
   (b) 2.75 m
   (c) 3 m
   (d) 3.75 m
4. Enclosed cable systems are most commonly used in which type of installation?
   (a) industrial
   (b) domestic
   (c) caravan
   (d) temporary
5. If enclosures are to be installed in cast in situ concrete, care must be taken to avoid
   (a) expansion
   (b) corrosion
   (c) bends and sets
   (d) using cpcs
6. Conduit and trunking, whilst being stored, must be
   (a) painted
   (b) correctly supported
   (c) on the floor
   (d) left uncovered
7. The tool used for for deburring a cut conduit prior to installation is called a
   (a) reamer
   (b) rammer
   (c) roamer
   (d) reaper
8. Cables installed in trunking and conduit systems are provided with
   (a) thick mechanical protection
   (b) thin mechanical protection
   (c) no mechanical protection
   (d) regular mechanical protection
9. Steel and PVC are used for the manufacture of conduit. One other material which has been widely used is
   (a) rubber
   (b) concrete
   (c) aluminium
   (d) brass
10. The effect of current flow through a ferrous metal enclosure produces
    (a) a magnetic field which aids current flow
    (b) a magnetic field which has no effect on current flow
    (c) a magnetic field which opposes current flow
    (d) no effect

Answer grid

1	a	b	c	d		6	a	b	c	d
2	a	b	c	d		7	a	b	c	d
3	a	b	c	d		8	a	b	c	d
4	a	b	c	d		9	a	b	c	d
5	a	b	c	d		10	a	b	c	d

5

Conduit Systems

You will need to have available for reference a current copy of IEE Guidance Notes 1 in order to complete the exercises within this chapter.

Answer the following questions to show you have understood the previous chapter.

The type of cable that is usually installed in enclosed cable systems is _____

The most likely location you will find an enclosed system is in _____

Why would plastic conduit be installed in preference to steel conduit when it is encased in concrete?

On completion of this chapter you should be able to:

◆ relate the types of steel conduit to applications
◆ recognise the need for minimum bending radius
◆ describe the methods of fixing conduit to surfaces
◆ recognise typical conduit fittings
◆ state some of the advantages and disadvantages of using rigid PVC conduit
◆ recognise the need to provide a separate cpc for each circuit contained in a rigid PVC conduit system
◆ recognise the effects of heat on rigid PVC conduit and the necessary precautions to take when assembling
◆ recognise the need to take precautions when using solvent adhesives
◆ state the maximum distance between supports
◆ list the materials that flexible conduits are manufactured from
◆ recognise that all flexible conduit systems must have a separate cpc installed
◆ describe the method of terminating flexible conduits
◆ identify different systems of mechanical protection for sheathed cable systems
◆ recognise the factors that must be taken into consideration when installing channelling
◆ state the standard length channel is supplied in
◆ recognise the requirements when cable is installed flush in a building structure
◆ complete the revision exercise at the beginning of the next chapter

Part 1

In this part we shall look at the various types of conduit that are used, the types of fittings and fixings associated with them and the regulations regarding their installation.

Figure 5.1   Conduit-bending machine

Remember
Conduit is sized according to its outside diameter.

20 mm

Types of steel conduit

This is available in round section as two main types, welded seam and solid drawn.

The welded seam type is the one most commonly used and is constructed from a flat steel plate rolled into a tube. The two butt edges are then welded together as shown in Figure 5.2.

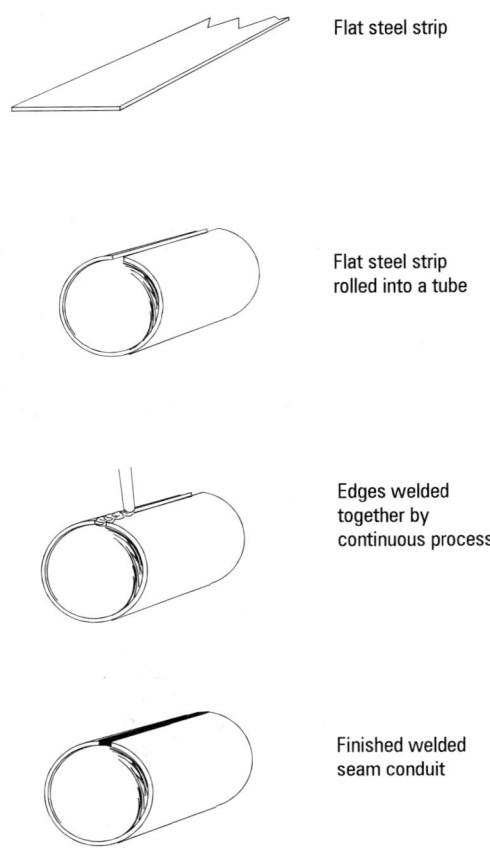

Flat steel strip

Flat steel strip rolled into a tube

Edges welded together by continuous process

Finished welded seam conduit

Figure 5.2   The construction of welded seam steel conduit

The outside edge is then ground smooth to complete the construction of the tube. If you look carefully at the inside of the tube you can see the weld running down the tube. This type of conduit is used for most general installation work as it is the cheapest of the two and is ideally suited to this purpose.

Solid drawn conduit is made from a large-diameter cast steel tube which is then drawn through a series of dies which reduce the diameter of the tube and increase its length. This process is continued until the required outside diameter is achieved.

Solid drawn conduit is less common as it is more expensive, but it is used where an installation has to be carried out in a flammable gas or explosive atmosphere to ensure that no gas or flammable vapour can make its way into the conduit, as could occur through a pinhole in the weld.

Finishes of steel conduit

Both of these types of conduit are available in a variety of finishes. The most common of these is **black enamel**. For this finish the tube is cleaned and then sprayed with black enamel paint. This finish is for general-purpose use in areas where it will not be exposed to adverse conditions such as moisture and condensation. The finish may be damaged during the process of cutting and threading, so once the conduit system is assembled any damaged surface or exposed threads should be repainted with black enamel paint.

If conduit is to be installed in an area where it will be subjected to damp, moisture or rain we must use a better suited finish. For this we use **galvanised conduit**, which means the conduit has been cleaned and then dipped in hot zinc. This adheres to the steel and provides a rust-resistant finish. If this is damaged it must be repaired using a rust-inhibiting paint. As the surface is removed during the threading process it is important that any exposed threads are also treated to prevent corrosion.

It may be that we have to install conduit in a very hostile environment, such as a sea front, where it will be exposed to salt water spray, which of course is highly corrosive. To combat this we can use a **sherardized conduit**, a process which involves coating the conduit whilst it is hot with zinc dust. This results in the steel becoming impregnated with zinc so that it does not flake off or crack. This produces a strong corrosion inhibiting surface, but again we must protect any exposed threads on completion.

So these are the types and finishes of steel conduit that are commonly used. Let us now take a look at the way in which a steel conduit system is put together.

Joining steel conduit

This type of system is screwed together and a thread may be cut onto a plain piece of conduit using a set of stocks and dies, as shown in Figures 5.3 and 5.4.

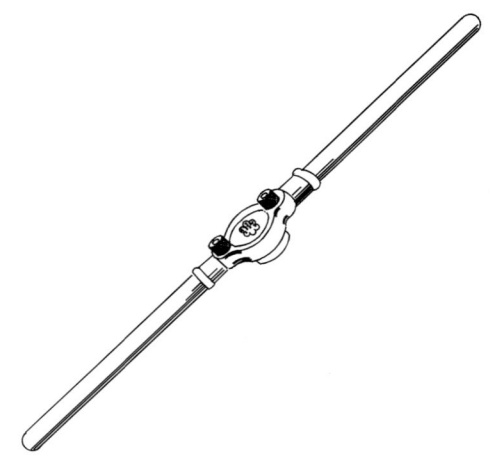

Figure 5.3   Stocks and dies

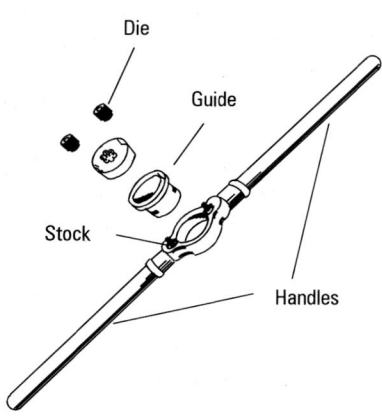

Figure 5.4   Parts of the set of stocks and dies

The conduit is held in a pipe vice (Figure 5.5) and cut to length and square using a hacksaw and file if necessary. All rough edges and burrs must be reamed or filed off, as cables will have to be pulled through the inside.

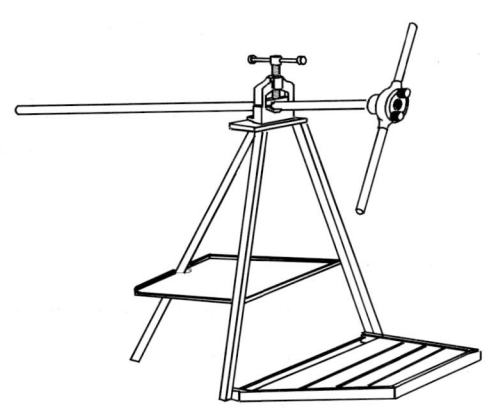

Figure 5.5   Pipe vice

A thread is then cut onto the conduit using the stocks and dies.

To join two lengths of conduit together we use a "coupler" (Figure 5.6). This is a short length of pipe with a thread cut on the inside. Each piece of conduit has a thread cut on it half the length of the coupler and the two pieces are screwed together until the ends butt together (Figure 5.7).

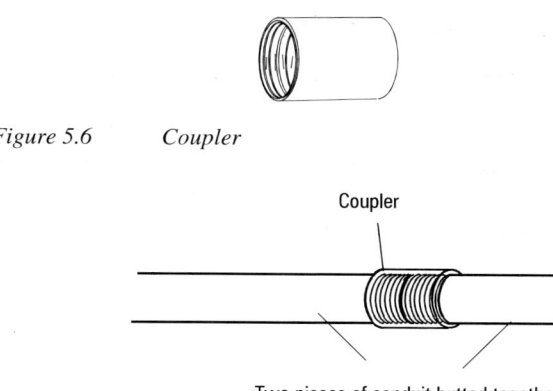

Figure 5.6   Coupler

Figure 5.7

Now our conduit will not always be required in straight lengths, and we may have to put a bend in the conduit to change direction or overcome obstacles. This is done using a bending machine (Figure 5.8) or wooden setting block (Figure 5.9). As it is difficult to set conduit and achieve the required bend with a smooth radius using a wooden setting block, the bending machine is the method generally used by industry.

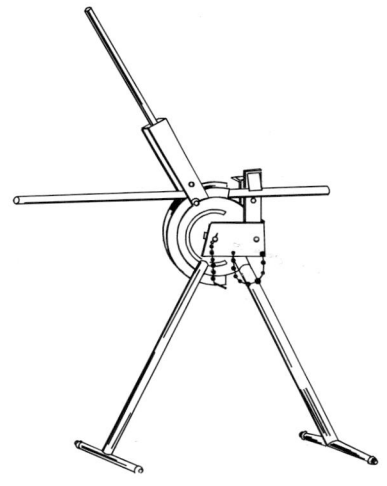

Figure 5.8   Bending machine

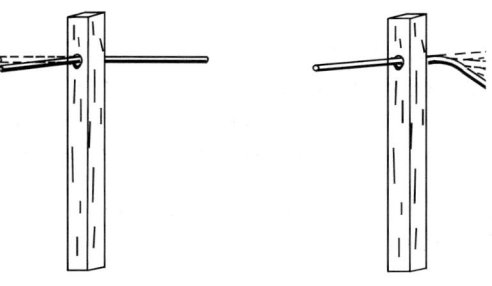

Figure 5.9   Wooden setting block

We can bend and join our conduit to the desired shape, but remember that we have to install the cables into the conduit, and Regulations require that we completely erect the conduit system before cables are installed. When the conduit is terminated into boxes the bushes should be tightened so that continuity is maintained. A bush spanner (Figure 5.10) is available to help with this.

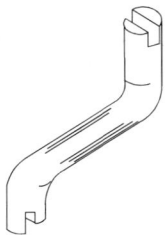

Figure 5.10   Bush spanner

Two problems then face us.

Firstly, we have to be able to get the cables into the conduit and out again at various points, for example lighting fittings.

Secondly, there is a limit to the length of conduit through which we can pull cables in one go. This will depend on the size of conduit and the size and number of cables. To overcome these problems we use "draw-in" boxes (conduit boxes).

Conduit boxes

Conduit boxes come in a variety of forms, and some of these are shown in Figure 5.11.

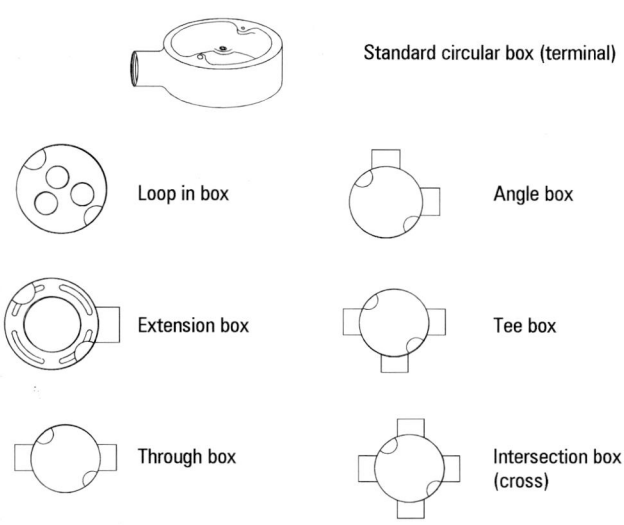

Standard circular box (terminal)

Loop in box

Angle box

Extension box

Tee box

Through box

Intersection box (cross)

The next group of boxes all have a spouted back outlet.

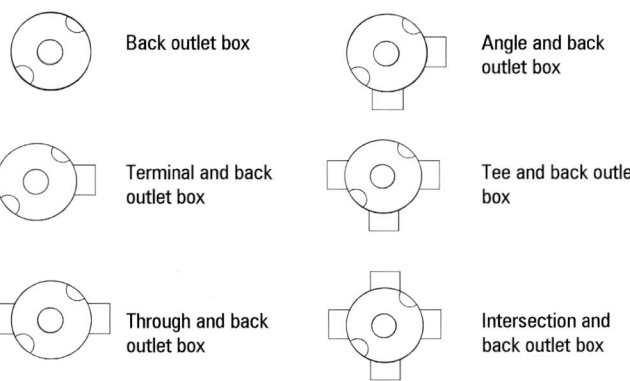

Back outlet box

Angle and back outlet box

Terminal and back outlet box

Tee and back outlet box

Through and back outlet box

Intersection and back outlet box

The following boxes all have tangent entries.

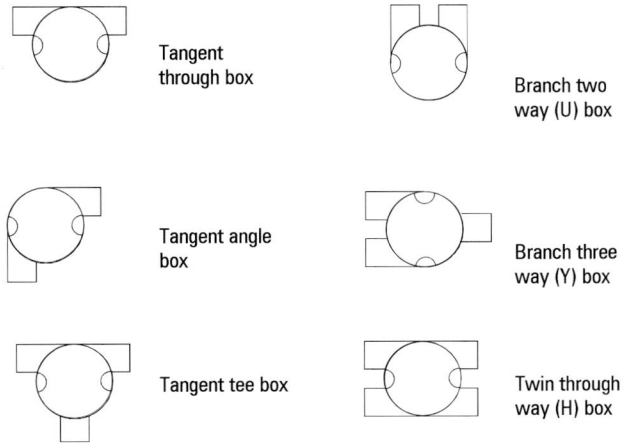

Tangent through box

Branch two way (U) box

Tangent angle box

Branch three way (Y) box

Tangent tee box

Twin through way (H) box

Figure 5.11

We may use these boxes in a variety of ways. The straightforward boxes have spouts with threaded entries, placed on the central axis of the circular box and these are used for most standard runs of conduit. In addition to these we have what is known as "tangent" boxes, where the spouts are arranged on the edges of the boxes. If we compare the angle boxes shown in Figure 5.12 with the tangent boxes shown in Figure 5.13 the difference is fairly obvious. The tangent box would be used where we have to run conduits close in to corners or along the angle between the ceiling and walls. Its design saves us setting the conduit whenever we have to fit a box.

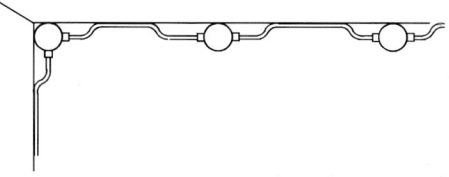

Figure 5.12   Installation using standard boxes

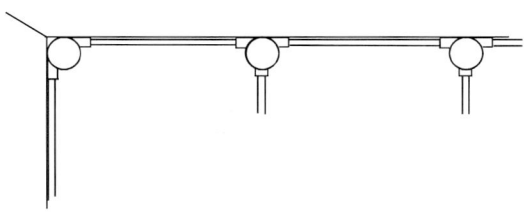

Figure 5.13   Installation using tangent boxes

Once the cables are installed these draw in boxes are sealed by fitting box lids so that the system is totally enclosed. If a high concentration of moisture is likely to be present then a rubber gasket should be fitted between the lid and the box to prevent water from entering the conduit system.

Where any system is not intended to be sealed, for example those not in flammable situations, they should be provided with drainage outlets at any point where moisture may collect. We can do this by fitting a tee box with an open spout pointing downwards at the lowest point on the system (Figure 5.14). This allows moisture to drain away if it should occur.

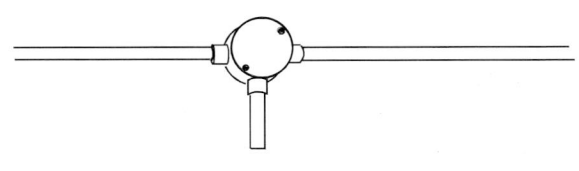

Figure 5.14   Drainage spout

In addition to the boxes shown in Figure 5.11 we also have other fittings, such as solid and inspection bends, elbows and tees. Some of these are shown in Figures 5.15–5.18.

Figure 5.15   Solid elbow

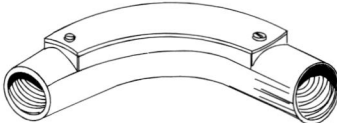

Figure 5.16   Inspection bend

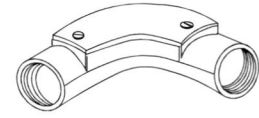

Figure 5.17   Inspection elbow

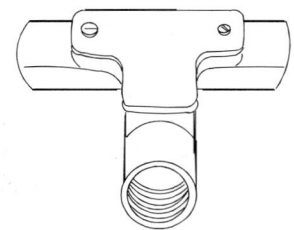

Figure 5.18   Inspection tee

As the solid fittings have no lid and are of a smaller radius than the minimum allowed, the use of these solid fittings is restricted.

Typically we could use solid fittings immediately behind a luminaire or an outlet box, as shown in Figure 5.19. Remember that BS 7671 requires that we ensure that conductors or insulation are not damaged during the installation of cables in conduit systems. Providing we fulfil these requirements we may use solid fittings in locations where space is restricted and yet changes of direction and so on are needed.

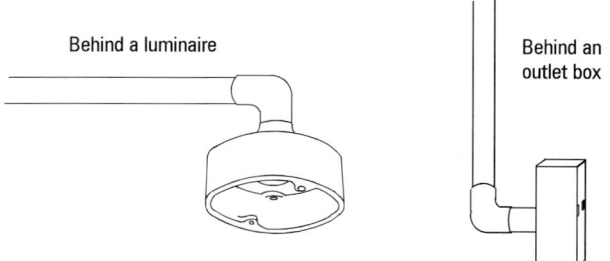

Figure 5.19   Typical uses for solid fittings

Bending radius

The old industry standard for the minimum bending radius of conduit was accepted as 2.5 times the outside diameter of the conduit, and this is still widely accepted as good practice (Figure 5.20).

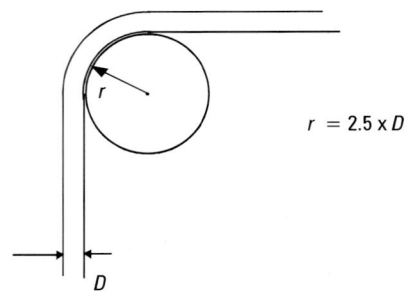

$r = 2.5 \times D$

Figure 5.20

Example:

If a 20 mm conduit is used then the minimum radius will be:

$2.5 \times 20 = \textbf{50 mm}$

Try this

Using the minimum bending radius as being 2.5 times the outside diameter of the conduit what are the minimum bending radii of the following conduits?

1. 16 mm

2. 25 mm

3. 32 mm

Points to remember ◀ — — — — — — — — — — — — —

Steel conduit is available in welded seam or _____ construction.

The finishes for steel conduit are black _____, galvanised and _____.

A steel conduit system is screwed together using _____ and a variety of conduit boxes.

The construction of a conduit system should be such that cables may be drawn into the system without damage to the conductors or insulation. To achieve this we may need to install additional draw in points.

Conduit boxes come in a variety of forms and the selection of the appropriate draw in box for each application will ensure a neat system with sufficient draw in points to prevent damage to cables during installation.

The minimum recommended bending radius for conduit is _____ × the outside diameter of the conduit.

Self-assessment multi-choice questions.
Circle the correct answers in the grid below.

1. The type of conduit used for general purpose installations is
   (a) solid drawn
   (b) galvanised
   (c) welded seam
   (d) sherardized
2. The type of conduit used in explosive gas situations is
   (a) solid drawn
   (b) galvanised
   (c) welded seam
   (d) sherardized
3. The conduit finish that is used in damp situations is
   (a) black enamel
   (b) galvanised
   (c) sherardized
   (d) PVC coated
4. The minimum recommended bending radius of a 25 mm conduit is
   (a) 75 mm
   (b) 67.5 mm
   (c) 62.5 mm
   (d) 57.5 mm
5. Conduit systems must have sufficient draw in points to ensure that
   (a) the layout is symmetrical
   (b) only one bend is placed between them
   (c) cables are not damaged during their installation
   (d) a maximum of two lengths of conduit exists between them

Answer grid

1	a	b	c	d
2	a	b	c	d
3	a	b	c	d
4	a	b	c	d
5	a	b	c	d

Part 2

Conduit fixings

We have looked at the component parts of a steel conduit system, but we must consider the types of fixings that are used to secure our conduit. If conduit is to be installed flush, buried in plaster, then a "crampet" is generally used to secure the conduit until the finish is applied. It is a simple single "nail" type fixing which holds the conduit adequately and is relatively cheap. A typical crampet is shown in Figure 5.21.

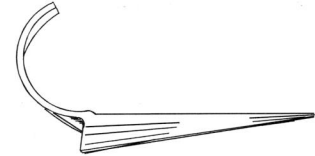

Figure 5.21   Crampet

For surface mounting we need something that is a little more robust and more convenient to use, allowing some adjustment or the ability to fit and remove conduit whilst the system is being erected. For this we can use a variety of fixings.

Saddles

The majority of surface conduits are fixed by using one of the following types of saddle fitting.

The plain saddle (Figure 5.22), which consists basically of a shaped steel strip. Two screws are installed into the surface on which the conduit is run. These saddles give the advantage of being able to fix the conduit after it has been installed. However, it will not allow easy adjustment of position for minor misalignment. Plain saddles will also not make any allowance for irregularities in the surface, such as on brickwork.

Figure 5.22   Plain saddle

For convenience when conduits have to run in tight corners **half saddles** (Figure 5.23) are used. These are similar to plain saddles but have a single hole fixing.

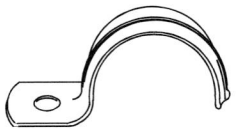

Figure 5.23   Half saddle

The spacer bar saddle (Figure 5.24). This consists of a flat steel plate which is fixed to the surface by a single countersunk screw. The conduit lays on this plate and a cover, similar to a plain saddle with two keyhole slots, attaches the conduit to the base using machine screws. This type of saddle overcomes the majority of the problems of the plain saddle allowing small movements to help alignment. Also, the bases may be fixed before the conduit is installed and the covers fitted and removed without affecting the fixing to the surface. It does allow for small irregularities in the surface, although larger irregularities will cause problems.

Figure 5.24   Spacer bar saddle

A distance saddle (Figure 5.25), as the name implies, distances the conduit from the surface, and in this case the space between the conduit and the surface to which it is fixed is around 6 mm. This overcomes the problem of uneven surfaces (Figure 5.26) and also saves the setting out of conduit at switches or sockets. However, they are more costly and do not usually have the facility for adjustment once installed and so require more accurate positioning.

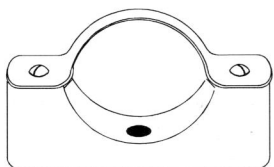

Figure 5.25   Distance saddle

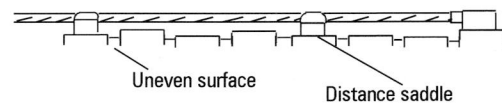

Uneven surface

Distance saddle

Figure 5.26

As we can see from Figure 5.25 the base is cast metal and the conduit is half set into the base. The result of this is that the conduit is a much closer fit in the saddle than in the case of the plain or spacer bar saddle.

The hospital saddle is the last type of saddle that we shall consider. These are specially designed to space the conduit away from the surface by about 12 mm and the base is specially shaped as shown in Figure 5.27 to fulfil the essential requirements of a hospital environment. These requirements are that:

• there is enough room between the conduit and the surface for cleaning and decorating
• there is minimal build-up of dust and dirt on the saddles

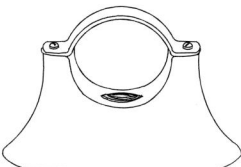

Figure 5.27   Hospital saddle

The curved base means that any dirt or dust rolls towards the front edge and down to the floor where it will be cleaned up during daily cleaning and so minimise the risk of germs breeding in dust residue.

It is common practice to install new hospital conduit systems flush in the structure to provide as clean a system as possible, but if conduit has to be installed on the surface then hospital type saddles should be used.

Girder clips

Many installations require us to fit conduits to roofs, voids and the like comprising a steel girder construction. Now, whilst it is possible to attach conduits using conventional saddles and drilling or cartridge fixing directly to the steel, in many instances this is structurally undesirable. Many specifications state that the structural steelwork is not to be drilled. A further consideration is that this method of installation is very time-consuming.

One method of overcoming this problem is by the use of girder clips. These are purpose-made grip fixings which allow a rapid fixing to steel construction which is structurally acceptable. These clips can be purchased for use with standard saddles or rapid fix "claw" or "Click Lock" type saddles. Typical examples are shown in Figures 5.28 and 5.29.

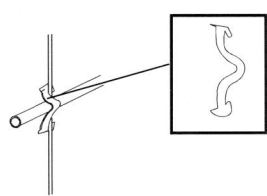

Figure 5.28   Conduit supported on rods

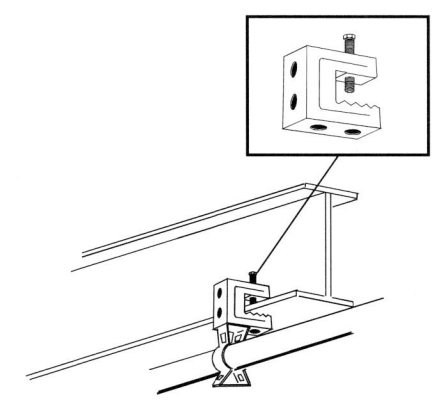

Figure 5.29   Beam clips and clamps

For single conduit runs these saddles provide a very quick installation method. If we have several conduits sharing the same route then clamp-type girder clips with outriggers can be fitted with a number of conduit fixings. This will allow us to produce a quick installation, and much of the support system can be constructed at ground level before erection. A bonus in using this method is that it allows us to ensure that all our conduits run parallel, giving a neat and professional finish. A typical example is shown in Figure 5.30.

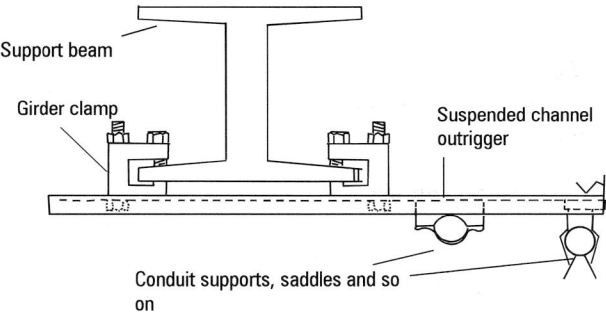

Support beam

Girder clamp

Suspended channel outrigger

Conduit supports, saddles and so on

Figure 5.30

Spacing of supports

The maximum distances between the supports (Figure 5.31) for our steel conduit system depend on

• the size of the conduit
• whether the conduit is run vertically or horizontally

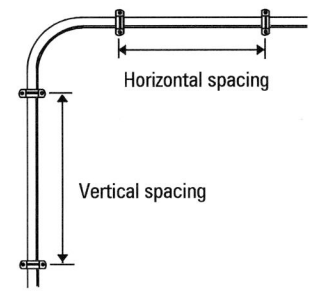

Horizontal spacing

Vertical spacing

Figure 5.31   Conduit supports

Try this

Using IEE Guidance Note 1 or an alternative complete the following table of examples of spacing of supports for conduit.

Table 5.1

Nominal size of conduit mm	Maximum distance between supports	
	Rigid metal	
	Horizontal m	Vertical m
< 16		
> 16 < 25		
> 25 < 40		
> 40		

Try this

Using manufacturers' catalogues compile a list of available conduit fixings, saddles etc, and identify any that are not included within the text of this module.

Points to remember ◄ — — — — — — — — — — — — —

What type of fixing is suitable to be used

(a) where the conduit is buried in plaster?

(b) where the surface to be fixed to is very uneven?

What are the requirements for fixings in a hospital environment?

If you are unable to drill directly into a steel girder, what would you suggest is the best method of fixing for the conduit?

What is the maximum distance apart for supports of a run of 20 mm conduit horizontally?

What is the maximum distance apart for supports of 32 mm conduit run vertically?

1. The best method of fixing a conduit to an uneven brick surface is by use of a
   (a) half saddle
   (b) crampet
   (c) open saddle
   (d) distance saddle
2. Conduit to be secured in a chase in a brick wall before plastering is best done with a
   (a) hospital saddle
   (b) crampet
   (c) spacer bar saddle
   (d) distance saddle
3. On completion of a black enamel conduit system any exposed threads should be
   (a) lubricated
   (b) galvanised
   (c) painted
   (d) shrouded
4. The box shown in Figure 5.32 is a

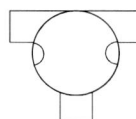

Figure 5.32

   (a) through box
   (b) angle box
   (c) tangent tee box
   (d) tee box
5. The saddle shown in Figure 5.33 is a

Figure 5.33

   (a) plain saddle
   (b) spacer bar saddle
   (c) distance saddle
   (d) hospital saddle

Answer grid

1	a	b	c	d
2	a	b	c	d
3	a	b	c	d
4	a	b	c	d
5	a	b	c	d

Part 3

Plastic conduit

In this part we shall look at the types of plastic conduit in common use, their associated fittings and fixings. We shall also see how a plastic conduit system is assembled and consider its applications.

Figure 5.34

Types of plastic conduit in common use

Rigid PVC (uPVC) conduit is available in circular section in the same sizes as steel conduit; that is, 16 mm, 20 mm, 25 mm and 32 mm. Larger sizes up to 50 mm are also available. Like our steel conduit there are two main types, these being "light gauge" and "heavy gauge", the difference between them being the thickness of the conduit wall. Now, remembering that conduit is measured and sized by its outside diameter, this must mean that the thicker the wall of the conduit the less internal space is available for cables. This, however, does not create a problem for our conduit capacity as the overall space reduction is very small. It is generally available in black or white in both heavy and light gauge.

Whilst plastic conduit is very robust it may need extra mechanical protection where it could be subjected to mechanical damage, for example at low levels or where heavy materials are moved around. Care also needs to be taken when bending and setting rigid PVC conduit, especially with the light gauge.

Light gauge rigid PVC conduit is best suited to areas where it
- is not likely to be mechanically abused
- does not need to have many bends or sets as it is not as robust as steel or heavy gauge rigid PVC conduit

For these reasons it is most commonly used where conduit is going to be out of sight, protected from mechanical impact and where gradual bends may be used.

59

Heavy gauge rigid PVC conduit is surprisingly robust and may be used in place of steel conduit for a great many applications. If care is taken it may be bent and set fairly easily and it is the most commonly used of the plastic conduits.

Some plastic conduit is marked as "High Impact" which indicates that it is suitable for use in situations where it may be subject to some mechanical abuse. Whilst this is not as strong as steel it is very robust. For example, it is not easily broken by the hammer impact should you miss hit a crampet fixing.

Crampet fixings are often used to secure plastic conduits in carcass installations. This is acceptable when the conduit is to be encased in cement or plaster. Care should be taken to ensure that, where there is a plaster finish, any metal, such as a crampet fixing, is painted to prevent corrosion. If this is not done the rust which forms will come to the surface as a brown stain and mark the decor.

During the installation care must be taken to make sure we do not damage the conduit.

Remember
A misguided hammer blow could split or shatter light gauge PVC conduits.

Advantages and disadvantages of rigid PVC conduit

Rigid PVC conduit, in general, has a number of advantages and as a result it has increased in popularity. We shall consider the merits of rigid PVC conduit and look at the problems that we may have to consider if we choose to use this system.

One of the main advantages for us as installers of a conduit system is that rigid PVC conduit is only 1/6th the weight of steel conduit and rather more flexible. This means that it is more easily handled, stored and transported, though we must remember that it does need more support than steel to prevent permanent distortion during transport and storage.

During installation of a concealed system it is possible to "flex" rigid PVC conduit around an obstacle making the installation of the conduit easier and quicker and give us a better route for pulling in the cables. An example of this is shown in Figures 5.35 and 5.36.

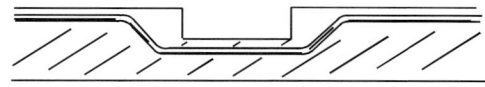

Figure 5.35   *Steel conduit in a concealed location using bends around an obstruction*

Figure 5.36   *Flexing plastic conduit around the same obstruction.*

The nature of PVC (**P**oly**V**inyl **C**hloride) gives us several other advantages:

- it is highly resistant to corrosion caused by water, acid and oxidising agents in concentrations that are likely to occur in domestic and general industrial installations (Figure 5.37)
- it shows little sign of deterioration after long periods of exposure out of doors (Figure 5.38)
- it is an excellent electrical insulator

So we can install the same type of conduit both inside and out, or bury it in plaster, without needing to take any precautions with regard to the finish.

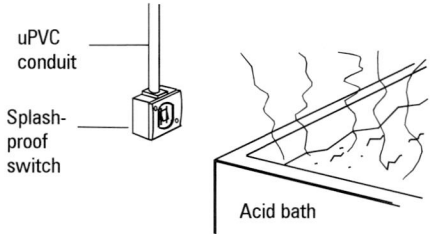

Figure 5.37   *Industrial installation*

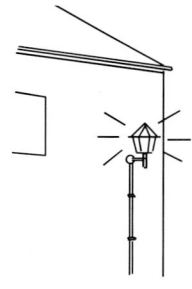

Figure 5.38   *External installation*

That's quite a list of advantages, but as always there are some drawbacks. We found out earlier that rigid PVC conduit does need extra mechanical protection where it is liable to mechanical damage, so that is the first minor drawback. The next problem that we must consider is that of a circuit protective conductor (cpc).

As PVC is a good insulator it follows that we cannot use the rigid PVC conduit as a cpc. So to overcome this we must install a separate cpc for each circuit we install. This will affect the number of circuits that can be installed in a given size of conduit.

Imagine that we are going to install six lighting circuits in a conduit system. If we use a steel conduit we may use the conduit itself as the cpc. In this case, each circuit will require one phase and one neutral conductor. So with six circuits we will have to install a total of 12 conductors (6 phase and 6 neutral).

The steel conduit would need to have a capacity suitable for 12 conductors (Figure 5.39).

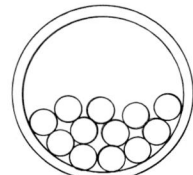

Figure 5.39   Steel conduit with 12 conductors

If the same installation is carried out using rigid PVC conduit we must install cpcs for the circuits, so our total number of conductors could be up to be 18 (6 phase, 6 neutral and 6 cpc). So our rigid PVC conduit could have to have a capacity for 18 conductors (Figure 5.40).

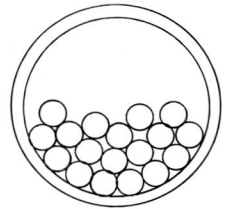

Figure 5.40   uPVC conduit with 18 conductors

In each case we have six circuits, but the rigid PVC conduit must have a greater capacity.

These cpcs must be insulated conductors of appropriate cross-sectional area and are coloured green and yellow, this being the accepted standard colour code for earthing conductors.

The other major problem with rigid PVC conduit is that of expansion and contraction with change of temperature. If your house is fitted with plastic guttering and downpipes you can often hear these making clicking and cracking noises during the evening after a hot day. This is caused by the plastic expanding during the day as it gets hot and then contracting when the sun goes down and the temperature falls. A similar thing happens to our rigid PVC conduit.

Now a change in temperature for most materials produces either change of volume of the material or change in pressure and often both. In the case of our rigid PVC conduit the change in volume of the material produces a change in the length of conduit. Some change in diameter will also occur, but this tends to be minimal. If the conduit cannot move this expansion will cause the conduit to buckle.

To allow for expansion in our conduit we use special expansion couplers to allow our conduit to expand and contract without exposing the conductors. We shall look at the construction of these and how we make the joints later when we go through the construction of the rigid PVC conduit system.

Remember
We must install cpcs for circuits in rigid PVC conduit installations.
Heat is produced by cables carrying current and this heat production stops when the current is switched off.

Constructing the rigid PVC conduit system

As for steel conduit, BS 7671 requires that the rigid PVC conduit systems buried in the building structure must be completed before cables are installed.

Bends and sets in rigid PVC conduit may be installed simply by using a bending spring or a former of the correct radius. BS 7671 requires that cables contained in a conduit are not bent so sharply as to cause damage or be damaged during the drawing-in process, but does not stipulate a minimum internal bending radius. As an example of good practice, the bending radius should not be less than 2.5 times the outside diameter. However, when installing PVC cables into a rigid PVC conduit the two similar materials tend to bind together, making it difficult to install conductors without damage to either conductor insulation or the conduit. It is common therefore to use a radius of around 4 times the outside diameter to reduce this problem (Figure 5.41).

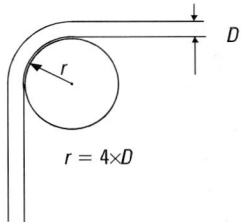

$$r = 4 \times D$$

Figure 5.41

A bending spring is fitted inside the conduit prior to installing the bend (Figure 5.42). It may be necessary to apply some heat to the conduit before bending, especially in cold weather, to prevent the wall from splitting. This is best done by using friction – rub a piece of rag up and down the conduit in the area where it is to be bent.

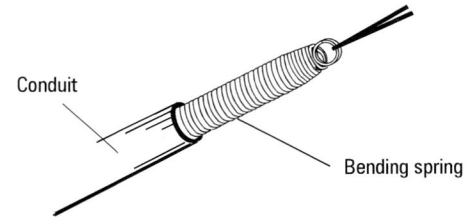

Figure 5.42   The bending spring inside the conduit

We can buy preformed 90° bends for all sizes of conduit to both 2.5 × outside diameter and 4 × outside diameter, but for the smaller sizes it is as easy and cheaper to use the bending spring. For the larger sizes where bending is more difficult it is often quicker and easier to use preformed bends.

When jointing the steel conduit the system was screwed together. With our plastic conduit we simply glue the fittings and conduit together using a variety of adhesives depending on the type of joint required. The fittings used are similar to those used for steel conduit with one or two extras. Some examples are shown in Figure 5.43.

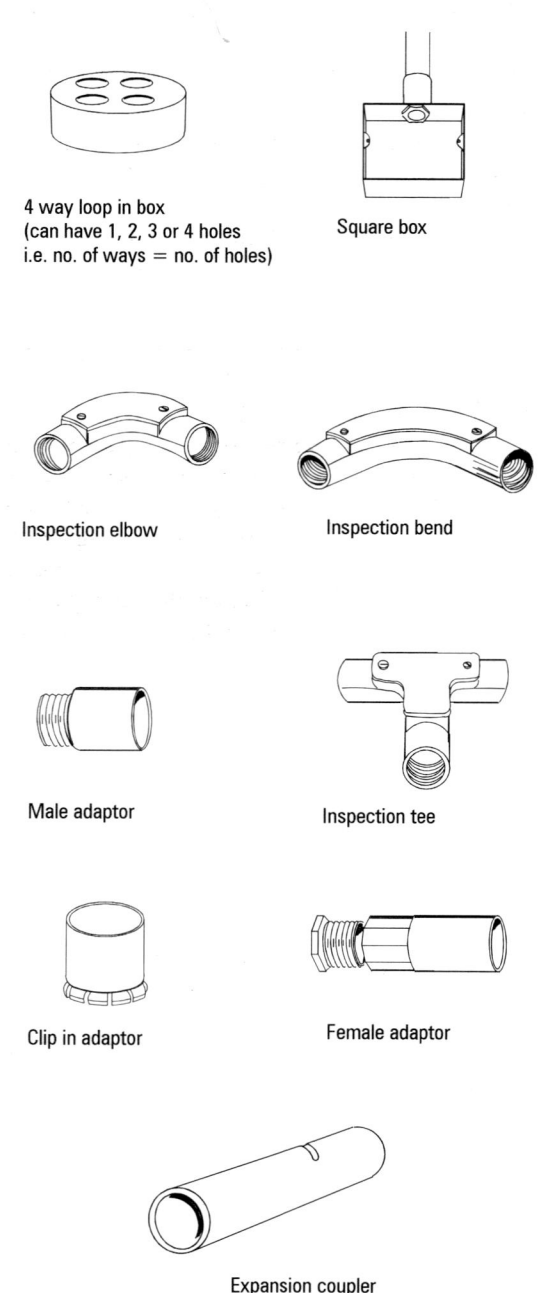

4 way loop in box
(can have 1, 2, 3 or 4 holes
i.e. no. of ways = no. of holes)

Square box

Inspection elbow

Inspection bend

Male adaptor

Inspection tee

Clip in adaptor

Female adaptor

Expansion coupler

Figure 5.43   Plastic conduit fittings

The conduit is simply pushed into the entries or couplers, and it is a common practice to assemble the system in stages and glue each stage when it is complete to allow fittings to be removed for adjustment or replacement. Once the first section is complete each joint is then glued together.

For most general-purpose joints a solvent adhesive is used. As with all solvents, care must be taken during their use. They should be applied in a well-ventilated environment and care must be taken to avoid inhaling the fumes they give off. It is also important to avoid contact with eyes, and contact with skin should be avoided whenever possible.

 If adhesive is being applied at high level, goggles or protective glasses should always be worn.

To assist us in this, most tins of adhesive come with a small brush attached to the screw-on lid (Figure 5.44), and it should always be used to apply the adhesive.

SOLVENT ADHESIVE

Figure 5.44

Once this type of joint is made it will not allow any movement or alteration, so it is important that we ensure that the fittings are correctly located before the adhesive sets.

We found out earlier that some allowance must be made for expansion and contraction due to change in temperature. These joints require the use of an expansion coupling. This is shown in Figure 5.45. The rate of expansion of PVC conduit is approximately 5 times that of steel.

Expansion coupling

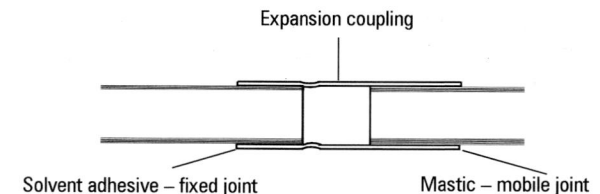

Solvent adhesive – fixed joint

Mastic – mobile joint

Figure 5.45

As we can see, the coupling has one standard length end and one much longer end. The standard end is glued to the conduit with solvent, whilst we joint the longer end using a mastic which allows the conduit to move. The conduit should not go right into the coupler on the free end. This will allow for expansion. If the conduit is allowed to expand and contract then the cables should not be tight in the conduit, but some slack should be allowed at draw-in points and termination points to prevent the conductors from being strained (Figures 5.46 and 5.47).

Figure 5.46   Draw-in point

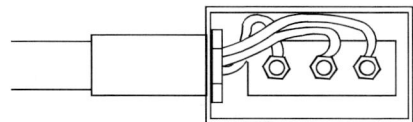

Figure 5.47   Termination point

In addition to allowing for expansion and contraction we may only install rigid PVC conduit in areas where it will not be subject to temperatures outside its permitted operating range. For ordinary rigid PVC conduit this is between –5 °C and 60 °C.

It may be recommended that boxes are used to provide support for luminaires and the like provided they can support the weight at the temperature that is likely to be present. For the temperature range given that the maximum weight that should be supported by each box is 3 kg.

We may support a 4 kg luminaire by suspending it from two or more boxes, as shown in Figures 5.48 and 5.49.

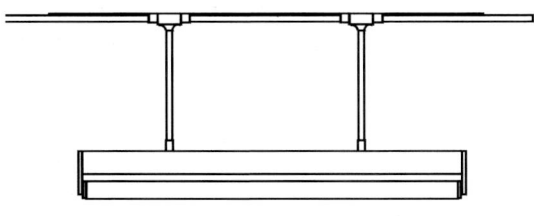

Figure 5.48   Luminaire suspended from two boxes

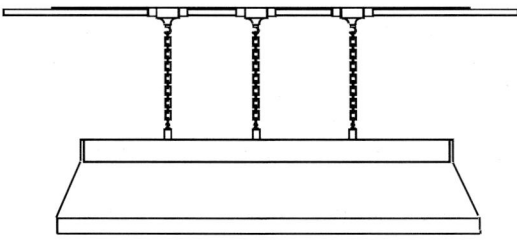

Figure 5.49   Luminaire suspended from three boxes

The final jointing cement we shall mention is a filler type cement. This is used to provide a watertight seal on conduits where they may be subjected to large amounts of condensation or out of doors where water may lie on joints after rain.

Remember
The temperature range for ordinary rigid PVC conduit is between –5 °C and + 60 °C
Within this range the maximum weight that should be supported by each box is 3 kg.

Rigid PVC conduit fixings

Rigid PVC conduit fixings are shown in Figure 5.50, and we can see that the plain saddle and spacer bar saddle are similar to those used for steel conduit, only they are made of plastic. In addition we have the "P" clip and the spring clip type of saddle. The spring clip is similar to the type of clip used to support water pipes, and once fixed to the surface the conduit is simply pressed home into the saddle, where it is held by the tension created.

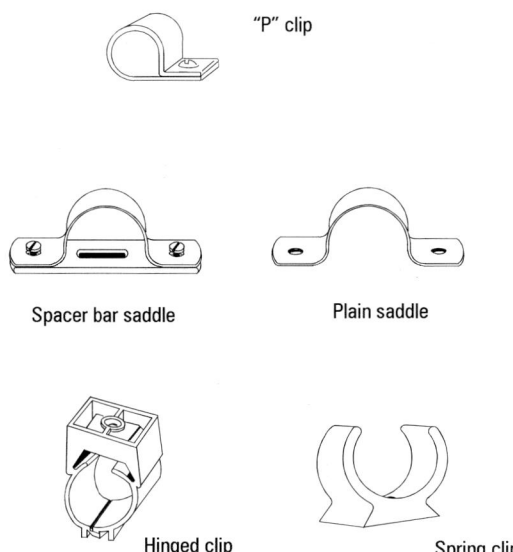

Figure 5.50   Rigid PVC conduit fixings

The clip fixing saddle is fixed to the surface, the conduit is placed in position and the hinged cover is clipped into place. This has the added advantages of

- the conduit cannot become dislodged due to movement
- it provides a secure fixing. Like the spacer bar or plain saddle and with a simple clip saddle there are less screws to be done up.

These fixings are also available in black or white and of course the appropriate colour should be used to match the conduit that we use.

Again we need to refer to the maximum distance between supports. In addition to manufacturers' advice, IEE Guidance Note 1 provides a table of the maximum distances between supports for conduit. If these distances for plastic conduit are compared with those for steel conduit we can see that for sizes above 16 mm the distances between fixings are reduced. This is because plastic conduit is more flexible than steel and so requires more support. For example the vertical distance for 20 mm steel conduit is given as 2 m between fixings and for PVC 1.75 m. The horizontal distance for 20 mm steel conduit is 1.75 m and for PVC 1.5 m. Supports should be positioned within 300 mm of bends or fittings.

If bends are installed using a spring or former it is often good practice to place saddles close to the bend, as shown in Figure 5.51.

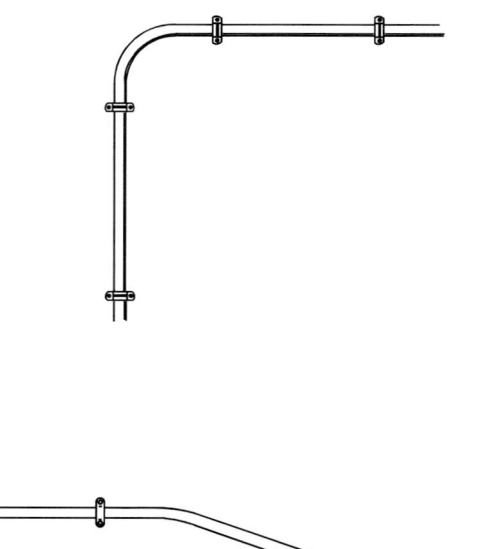

Figure 5.51   Place saddles close to bend or set

We do this because plastic conduit has a tendency to straighten after a period of time and so change the bend or set. By placing saddles close to these we can prevent this happening.

Try this

From the locations below select those that you would think would be suitable for using rigid PVC conduit.

Yes ✔ No ✘

1. A lighting circuit in a freezer store room.

2. To supply a light mounted on an outside wall.

3. To a control panel in a boiler room.

4. Socket outlet circuits, conduit beneath a concrete floor cast *in situ*.

5. A loop in lighting system in a large roof space with limited access and headroom.

6. Circuits running through a wash down area.

7. To supply flood lighting 3 m above an indoor swimming pool.

8. A socket outlet system in a motor vehicle repair workshop at low level.

9. A vertical conduit run with fixings only possible at 2 m intervals in 25 mm conduit.

10. To span a 2 m gap between two buildings at high level.

Plastic conduit is available in light and heavy gauge; both types are supplied in _____ or white.

Plastic conduit systems are joined by using a variety of adhesives, and care must be taken in their use as these give off _____ fumes until they are set.

An allowance must be made in a plastic conduit system for expansion and _____ due to changes in temperature.

If the temperature range of standard rigid PVC conduit is –5 °C to +_____ °C, within this range the maximum weight that should be supported by outlet boxes is _____ kg.

A separate cpc must be installed for each circuit contained in a plastic conduit system.

Try this

A lighting fitting with a weight of 25 kg is to be suspended above a snooker table. How many rigid PVC conduit outlet boxes would you need to install to support the fitting if they were to carry the whole weight of the fitting?

If the size of the fitting is 2 m × 1.5 m, sketch a layout in the space below showing how you would position the conduit boxes to best support the fitting.

Note: you may use more than the minimum number in order to achieve a good layout.

Self-assessment multi-choice questions

Circle the correct answers in the grid below.

1. Rigid PVC conduit is sized by its
   (a) external diameter
   (b) internal diameter
   (c) internal radius
   (d) external radius

2. The temperature range for rigid PVC conduit is
   (a) 0 °C to 65 °C
   (b) −10 °C to 65 °C
   (c) −5 °C to 60 °C
   (d) 5 °C to 60 °C

3. Provision must be made in a plastic conduit installation to allow for
   (a) vibration
   (b) expansion and contraction
   (c) installation of further conductors
   (d) length of runs

4. One of the advantages of plastic conduit is
   (a) once set it readily retains its shape
   (b) it is unaffected by temperature
   (c) it is only ⅙th of the weight of steel
   (d) it may be used as a cpc

5. The maximum weight that should be suspended from a plastic conduit box is
   (a) 2 kg
   (b) 3 kg
   (c) 4 kg
   (d) 5. kg

6. The fitting shown in Figure 5.52 is a

Figure 5.52

   (a) male threaded adaptor
   (b) female threaded adaptor
   (c) coupler
   (d) sleeved coupler

7. The fitting shown in Figure 5.53 is a

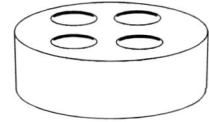

Figure 5.53

   (a) through box
   (b) 3 way box
   (c) loop in box
   (d) adaptor

8. Rigid PVC conduit is joined together by
   (a) interference fit
   (b) adhesives
   (c) screw threads
   (d) clip on fittings

9. The maximum distance between supports for a 20 mm vertical run of rigid PVC conduit is
   (a) 1.0 metres
   (b) 1.5 metres
   (c) 1.75 metres
   (d) 2.0 metres

10. To assist with the installation of conductors we often use bends with a radius of
    (a) 2 × outside diameter
    (b) 3 × outside diameter
    (c) 4 × outside diameter
    (d) 6 × outside diameter

Answer grid

1	a	b	c	d	6	a	b	c	d
2	a	b	c	d	7	a	b	c	d
3	a	b	c	d	8	a	b	c	d
4	a	b	c	d	9	a	b	c	d
5	a	b	c	d	10	a	b	c	d

Part 4

Alternative conduit systems and channelling

So far we have considered rigid conduit systems in both steel and plastic. In Part 4 we shall deal with the use of flexible conduits and their function, construction and the factors governing their use.

Our rigid conduit systems are firmly fixed to the structure of the building and are not designed to take movement or vibration, as this could loosen connections and may result in the conduit being fractured. Now, if we have to connect machinery, for example an electric motor, to our rigid system, then some vibration is likely and we may need to allow some movement for adjustment of drive belts and the like.

We overcome these problems by making the connection between the machine and the rigid system with a flexible conduit. This, as its name implies, is designed to move and can withstand any vibration that occurs. This flexible conduit comes in a galvanised steel finish and is available in a number of forms.

Flexible steel conduit

This is constructed in the form of a steel spiral with the edges interlinked to form a continuous tube, as shown in Figure 5.54. It may be cut to the desired length by cutting through one section and then "unscrewing" the two pieces away from each other as shown.

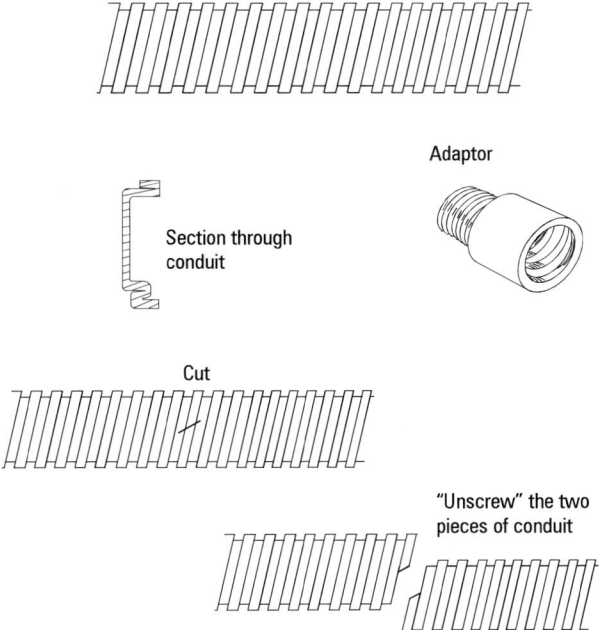

Figure 5.54

Once we have cut our conduit to the desired length a gland or "adaptor" is screwed onto the spiral at each end to enable us to connect to the rigid conduit system, usually at a draw in box conveniently located, and to the equipment.

Although made of steel, this flexible conduit may not be used as a circuit protective conductor, so a cpc must be installed inside the conduit with the other conductors to provide our earth continuity. This is in fact the case with all flexible conduits regardless of their construction.

Fibre flexible conduit

This is constructed with a number of layers bonded together. This can best be explained by Figure 5.55.

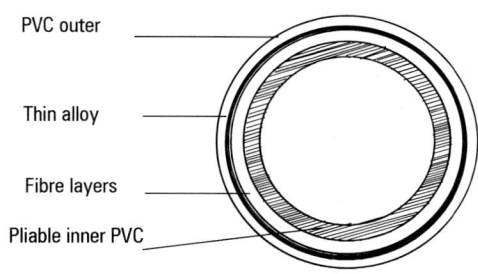

Figure 5.55

This type of flexible conduit is simply cut through with a junior hacksaw and the inside is cleaned. A gland is then fitted to each end. The conduit gland is shown in Figure 5.56.

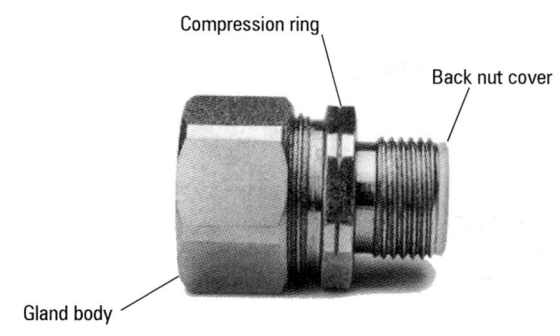

Figure 5.56

The gland body is screwed into the fitting first then the conduit is pushed home into the gland and the backnut is tightened down. This presses the compression ring and locks the conduit into position. This is probably the most common type of flexible conduit in use today.

It has the advantage of being somewhat stiff, so it can be moulded to a desired position which it will then retain. It is, however, affected by exposure to water and can break down physically should it get very wet.

There are some slight variations from one manufacturer to another, but the basic construction remains the same. Figure 5.57 shows a typical application for these two types of flexible conduit.

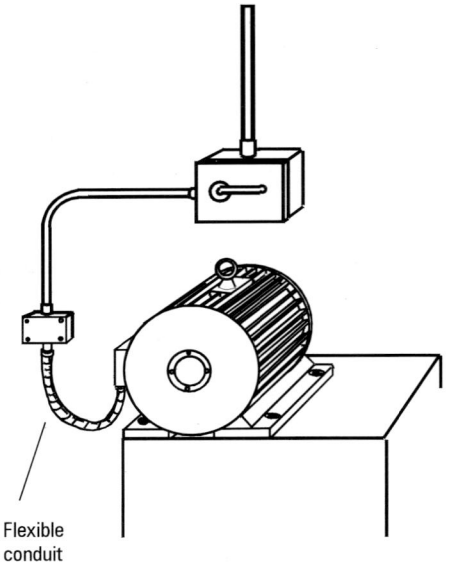

Flexible
conduit

Figure 5.57

Remember

A separate cpc must be installed in all flexible conduits to ensure earth continuity. The conduit itself may not be used as a cpc.

Pliable and corrugated plastic conduit

Whilst we are considering flexible conduits there is one other type that we should mention, and that is pliable PVC conduit (Figure 5.58).

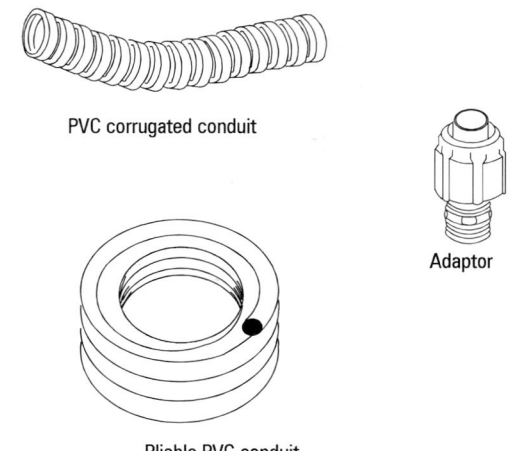

PVC corrugated conduit

Adaptor

Pliable PVC conduit

Figure 5.58

There are various types of pliable and corrugated PVC conduit and it is necessary to consult the manufacturer's catalogue to find their limitations. Figures 5.59 and 5.60 show a typical example of its use for a cooker outlet from a cooker panel in a stud partition wall. In such a location it is not practical to install a rigid conduit but the pliable or corrugated PVC conduit is suitable for this purpose.

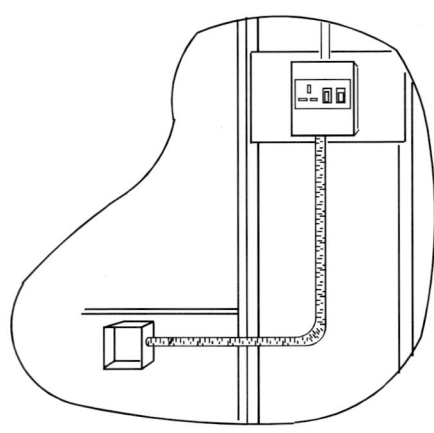

Figure 5.59   *Cutaway view of corrugated or pliable PVC conduit installation*

Figure 5.60   *Finished view of installation*

Remember

Flexible conduit forms the link between rigid systems, machines and so on absorbing vibration.

Flexible conduit may *not* be used as a cpc, and separate cpcs must be installed inside all flexible conduits.

Flexible conduits should not be so long as to require any support in their run.

Oval conduit

Oval conduit (Figures 5.61–5.64) is also used to provide extra mechanical protection for sheathed wiring, usually PVC/PVC. It is no longer widely used.

Oval conduit is usually supplied in 3 metre lengths and sizes between 12 mm and 32 mm measured across the largest axis. Whilst couplings are available to join straight lengths of oval conduit together no fittings are made to allow change of direction. It is not possible to bend oval conduit, although it may be flexed across its minor axis, and is therefore restricted to use in straight lengths.

At one time oval conduit was readily available in black enamel steel with an open seam down the centre of the major axis. Nowadays it is almost exclusively supplied in plastic.

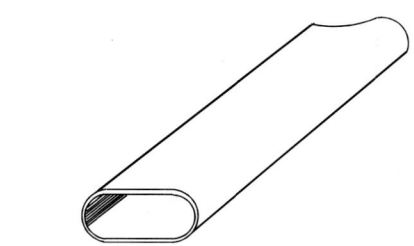

Figure 5.61   Oval conduit

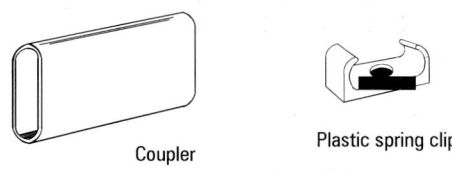

Coupler   Plastic spring clip

Figure 5.62   Oval conduit fittings

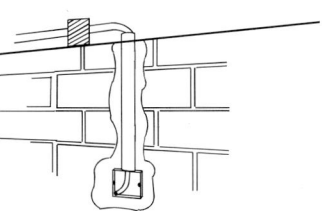

Figure 5.63   Oval conduit fitted flush

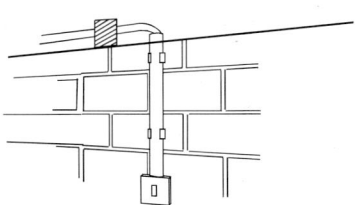

Figure 5.64   Oval conduit fitted on the surface

If oval conduit is to be fitted on the surface spring clips are available to attach it to the wall. However, as it is not always possible to terminate it into the moulded boxes neatly this use has been mostly given over to mini-trunking.

Should we use oval conduit flushed into the structure it will normally need to be chased into the brick or blockwork as, because of its depth, not enough plaster is applied to cover the conduit. Once installed it may be possible to draw in a new piece of cable through oval conduit, but its main function is to protect the sheathed cable from damage during the plastering process.

Channel

To achieve the same protection we can use a plastic or steel channel (Figure 5.65).

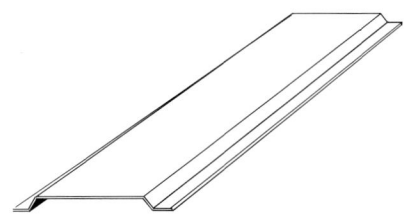

Figure 5.65   Plastic or steel channel

Channel is supplied in standard 2 metre lengths and widths from 12 mm to 38 mm and as we can see it is open backed. Channel will provide protection to sheathed cables from mechanical damage from the plaster's trowel during the plastering process and will also protect the PVC sheath from degeneration caused by the salts in the wet plaster. It is most commonly used in carcass wiring of domestic installations. It is fixed to the walls using galvanised nails (Figure 5.66) and is shallow enough to be covered by the plaster when this is applied.

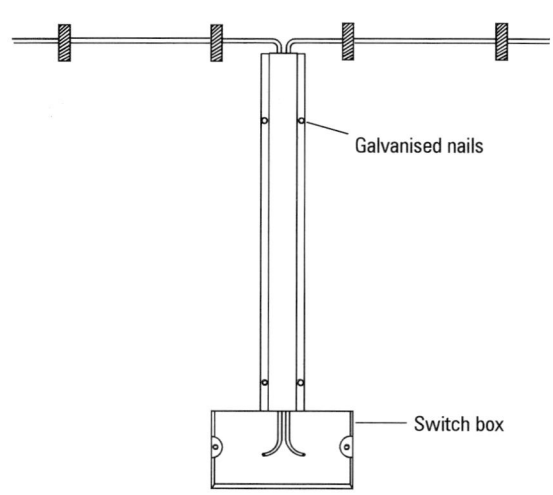

Figure 5.66

As there is not usually any chasing of masonry to be done, channel is a relatively cheap and quickly installed method of providing extra mechanical protection for our PVC/PVC cables. However, some care is needed during installation, so we must consider the following points:

- excess mortar between bricks or blocks must be removed to allow channel to lay flat against the structure
- cables must not be crossed over inside the channel
- cables must not be pinched either by the channel or the flanges
- channel must be fixed by the flanges; nails or pins must not be allowed to pierce the cables
- if metal channel is used sharp edges must be removed to prevent damage to the cables

If these points are kept in mind during installation our channel provides good protection at a relatively low cost. We must remember that it will not prevent anyone from drilling through cables, especially with a hammer drill, and because of this there are some requirements that we should observe when installing cables flush in the building structure:

- cables must be installed within 150 mm of the top of a wall or partition or within 150 mm of an angle formed by adjoining walls (Figure 5.67).
- if a cable is connected to a point or accessory on a wall or partition it may be installed outside of these zones only in a straight run either vertically or horizontally from the point (Figure 5.68).

If we cannot achieve either of the above conditions the cables used must either

- have mechanical protection sufficient to prevent penetration by nails, screws and the like
- be of concentric construction

These two conditions make it impractical to install PVC/PVC type of conductors as compliance will prove costly and time consuming. The best method is to avoid installing cables not vertical, horizontal or within the stated zones.

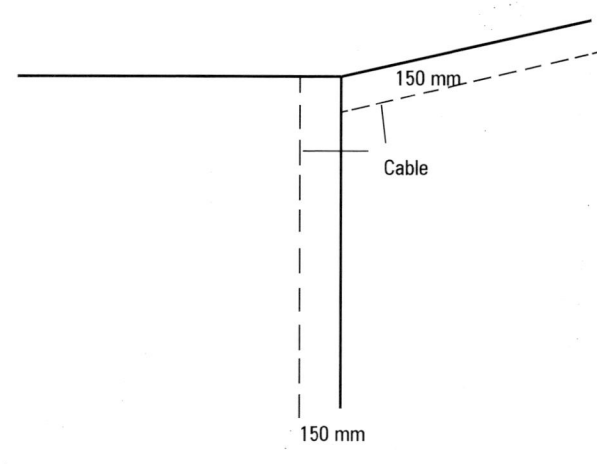

Figure 5.67   Cable to be installed within 150 mm of the top of the wall or within the angle between adjoining walls.

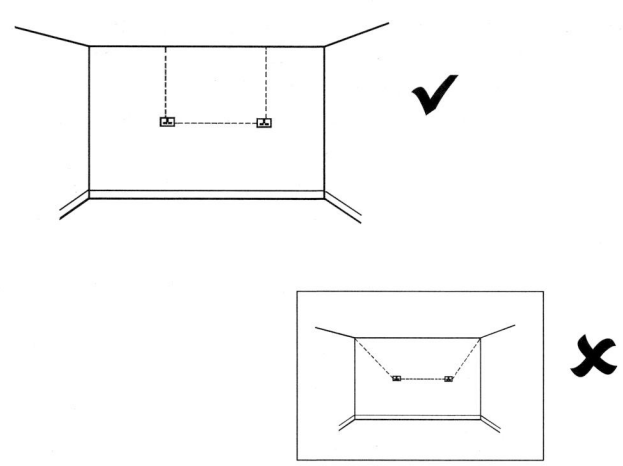

Figure 5.68   *Cables connected to an accessory must be run in a straight line either vertically or horizontally*

> **Remember**
>
> **Mechanical protection for sheathed cables may be provided by**
> - **oval trunking: restricted to straight runs and may be installed surface or flush**
> - **channelling: for flush installations and in either plastic or galvanised steel**
>
> **The installation of flush cables using channels is restricted by certain requirements to reduce the risk of damage to the cables.**

Points to remember ◄ – – – – – – – – – – – – –

All flexible conduits, regardless of their construction, must not be used as a circuit _____

_____

Flexible steel conduit is constructed in the form of a steel _____. Fibre flexible conduit is affected by exposure to _____

Oval conduit is also used to provide extra mechanical protection for sheathed wiring and is commonly used to protect wiring that is to be run _____ into the building structure. An alternative to using oval conduit in this situation is to use a plastic or steel channel.

Where cables are concealed within a wall they must be run within certain defined areas unless they are buried at a depth which need not exceed 50 mm.

Try this

If we are to install PVC/PVC cables select the most appropriate method of providing enclosures in the following installations.

1. Surface wiring to a number of socket outlets in a typing pool

2. Flush wiring in the carcass wiring of a house.

3. Drops to switches and sockets from cables installed above a suspended ceiling.

4. Surface wiring power and communications' cables to a midi computer installation.

Self-assessment multi-choice questions

Circle the correct answers in the grid below.

1. Galvanised steel flexible conduit
   (i) may be used as a cpc
   (ii) is resistant to corrosion
   (a) both statements (i) and (ii) are correct
   (b) only statement (i) is correct
   (c) only statement (ii) is correct
   (d) both statements (i) and (ii) are incorrect

2. The PVC-sheathed fibre-type construction of flexible conduit
   (i) retains the shape it is moulded to
   (ii) is suitable for locations where it may get wet
   (a) both statements (i) and (ii) are correct
   (b) only statement (i) is correct
   (c) only statement (ii) is correct
   (d) both statements (i) and (ii) are incorrect

3. Flexible conduit is terminated by means of
   (a) clips
   (b) clamps
   (c) screws
   (d) glands

4. The part indicated on the fibre type fitting shown in Figure 5.69 is:

Figure 5.69

   (a) body
   (b) backnut
   (c) filler
   (d) compression ring

5. The most likely type of flexible conduit used to supply a cooker outlet unit in a domestic stud partition wall is
   (a) fibre type
   (b) steel
   (c) pliable PVC
   (d) semi-rigid PVC

6. The largest standard size of oval conduit supplied as standard is
   (a) 16 mm
   (b) 20 mm
   (c) 25 mm
   (d) 32 mm

7. The smallest size of oval conduit supplied as standard is
   (a) 12 mm
   (b) 16 mm
   (c) 20 mm
   (d) 25 mm

8. Cable channel is supplied in standard lengths of
   (a) 3.0 metres
   (b) 2.5 metres
   (c) 2.0 metres
   (d) 1.8 metres

9. If cables are concealed in the building structure using channelling it should be run
   (i) Horizontal within 150 mm of the top of the wall
   (ii) Vertical within 150 mm of an angle formed by adjoining walls
   (a) statements (i) and (ii) are both correct
   (b) only statement (i) is correct
   (c) only statement (ii) is correct
   (d) statements (i) and (ii) are both incorrect

10. Which of the following statements is correct?
   (a) Oval conduit may be bent using a bending spring.
   (b) Oval conduit may be bent using a former.
   (c) Oval conduit may not be bent.
   (d) Oval conduit may be bent using elbows.

Answer grid

1	a	b	c	d		6.	a	b	c	d
2	a	b	c	d		7.	a	b	c	d
3	a	b	c	d		8	a	b	c	d
4	a	b	c	d		9	a	b	c	d
5	a	b	c	d		10	a	b	c	d

Progress Check

Circle the correct answers in the grid at the end of the multi-choice questions.

1. Which of the following is most commonly used as a cable conductor?
   (a) iron
   (b) brass
   (c) copper
   (d) silver

2. If a reel of copper had a resistance of 0.54 Ω a reel of similar aluminium cable would have a resistance of
   (a) 0.27 Ω
   (b) 0.57 Ω
   (c) 0.75 Ω
   (d) 0.9 Ω

3. If a reel of aluminium cable weighs 3 kg then a similar reel of copper cable would weigh approximately
   (a) 1.5 kg
   (b) 3 kg
   (c) 6 kg
   (d) 9 kg

4. The maximum operating temperature of silicon rubber insulation is
   (a) 55 °C
   (b) 85 °C
   (c) 150 °C
   (d) 250 °C

5. The reason for sealing the terminations of mineral insulated cable is to keep
   (a) moisture out of the cables insulation
   (b) the powder from falling out
   (c) the cable from drying out
   (d) vermin from eating the insulation

6. The layer of steel wires incorporated in an armoured cable
   (a) acts as a magnetic screen
   (b) protects the conductors from damage
   (c) provides an additional conductor when all the cores are in use
   (d) prevents external heat from reaching the conductors

7. The outer PVC sheath of MIMS cable is used to
   (a) act as a colour code for identification purposes
   (b) prevent electric shocks if the sheath goes live
   (c) prevent accidental contact between cables at different voltages
   (d) protect the outer sheath from the effects of electrolytic corrosion

8. The type of cable shown below is

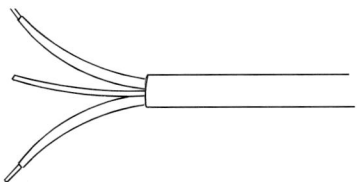

   (a) MIMS
   (b) PVC SWA
   (c) FP200
   (d) PVC-sheathed and insulated

9. Aluminium conductors of overhead cables are given additional strength by
   (a) incorporating a steel core amongst the strands
   (b) using solid rather than stranded construction
   (c) using a large number of fine strands
   (d) using triangular rather than round section

10. A cable having seven strands of 0.853 mm diameter has an overall cross-sectional area of
    (a) 2.5 mm^2
    (b) 4 mm^2
    (c) 6 mm^2
    (d) 10 mm^2

11. If a cable is replaced by one having twice the diameter the cross-sectional area will be
    (a) one and a half times greater
    (b) twice as much
    (c) four times as much
    (d) eight times as much

12. A single, wedge-shaped aluminium conductor, as shown below, forms exactly quarter of a circle. The dimension R is 8.74 mm. The cross-sectional area is

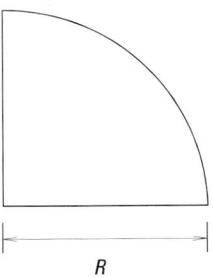

    R

    (a) 240 mm^2
    (b) 120 mm^2
    (c) 100 mm^2
    (d) 60 mm^2

13. Cables for high-voltage installations, when compared with those of low voltage need
    (a) thicker insulation
    (b) greater cross-sectional area
    (c) more strands
    (d) more cooling

14. Aluminium is used as a conductor in overhead cables in preference to copper because it is
    (a) lighter
    (b) stronger
    (c) a better conductor
    (d) easier to terminate
15. The normal method of fixing cable channel is with
    (a) spacer bar saddles
    (b) adhesives
    (c) galvanised nails
    (d) crampets
16. The wire used to support an overhead cable is called a
    (a) strainer
    (b) stress wire
    (c) hanger
    (d) catenary
17. A vertical conduit should incorporate means of support for cables at intervals not exceeding
    (a) 3 m
    (b) 5 m
    (c) 8 m
    (d) 10 m
18. The minimum internal bending radius for conduit systems should not be less than
    (a) 6 times the diameter of the conduit
    (b) 4 times the diameter of the conduit
    (c) 2.5 times the diameter of the conduit
    (d) that recommended for any of the cables to be contained
19. The maximum permissible distance between supports for horizontal steel trunking not exceeding 1500 mm^2 cross-sectional area is
    (a) 1.75 m
    (b) 2 m
    (c) 2.25 m
    (d) 2.5 m
20. A 230 V, 4.6 kW load with a power factor of 0.8 will draw a current of
    (a) 20 A
    (b) 23 A
    (c) 25 A
    (d) 30 A
21. The reference method which refers to cables clipped direct to the surface is Method
    (a) 1
    (b) 2
    (c) 3
    (d) 4

22. The conduit box shown below is a

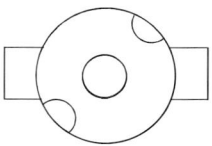

    (a) tee box
    (b) back outlet box
    (c) through box
    (d) back entry through box
23. The fixing device shown below is a

    (a) plain saddle
    (b) distance saddle
    (c) space bar saddle
    (d) crampet
24. To prevent PVC conduit collapsing when bent the conduit should be
    (a) heated up
    (b) bent cold
    (c) filled with hot water
    (d) internally reinforced by means of a spring
25. Where cables are concealed within a wall they must be run within certain defined areas unless they are buried at a depth which need not exceed
    (a) 10 mm
    (b) 20 mm
    (c) 35 mm
    (d) 50 mm

Answer grid

1	a b c d	16	a b c d
2	a b c d	17	a b c d
3	a b c d	18	a b c d
4	a b c d	19	a b c d
5	a b c d	20	a b c d
6	a b c d	21	a b c d
7	a b c d	22	a b c d
8	a b c d	23	a b c d
9	a b c d	24	a b c d
10	a b c d	25	a b c d
11	a b c d		
12	a b c d		
13	a b c d		
14	a b c d		
15	a b c d		

6

Trunking

You will need to have a copy of IEE Guidance Note 1 available for reference in order to complete the exercises within this chapter.

At the beginning of Chapter 4 the standard lengths and dimensions of trunking was stated. Complete the following sentences:

The standard length of trunking is _____.

The dimensions of the trunking range between

_____ × _____ and
_____ × _____

On completion of this chapter you should be able to:

- ◆ identify the different types of trunking available
- ◆ state the lengths and finishes of trunking as supplied
- ◆ recognise that trunking may not be used as a PEN conductor
- ◆ explain how to ensure electrical continuity when fabricating and joining trunking
- ◆ recognise that segregated trunking ensures separation of different categories of circuit
- ◆ select an appropriate trunking system for a given situation
- ◆ calculate the capacity of conduit or trunking
- ◆ identify which situations require the space factor to be used
- ◆ identify the three main types of installation that tables of cable factors are given for
- ◆ explain what is meant by "length of run" and how it can be reduced
- ◆ calculate the size of conduit or trunking required for given cables
- ◆ determine the maximum number of cables that can be installed in a given conduit or trunking
- ◆ complete the revision exercise at the beginning of the next chapter

Part 1

Having looked at the most common types of conduit in everyday use we shall now turn our attention to trunking. In this chapter we shall consider the types of trunking used, their manufacture and installation, and those made to perform a particular function. We will begin by looking at basic metal trunking.

Figure 6.1

Metal trunking

This is made from low-carbon sheet steel which is folded to form a rectangular section steel trough. A lid is then fitted to this to form the complete enclosure (Figure 6.2). It is made so that it may be removed to allow us to install cables. Manufacturers use different methods to attach this lid depending on their preference.

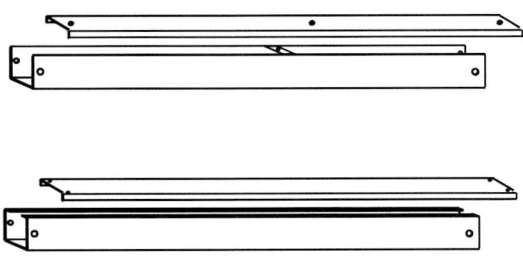

Figure 6.2

General-purpose trunking is normally supplied in 3 metre lengths, with sections from 38 mm × 33 mm to 225 mm × 100 mm. There are two basic finishes:

- galvanised
- grey enamel

and as for steel conduit the enamel type is used for general installations and the galvanised type is used where there is a risk of corrosion.

Why do we use trunking?

A conduit will carry a number of conductors sharing a common route and provide good mechanical protection, but it does have a limited capacity. If a large number of conductors are to be installed several conduits would have to be used, and crossovers would make the installation complex and unsightly (Figure 6.3).

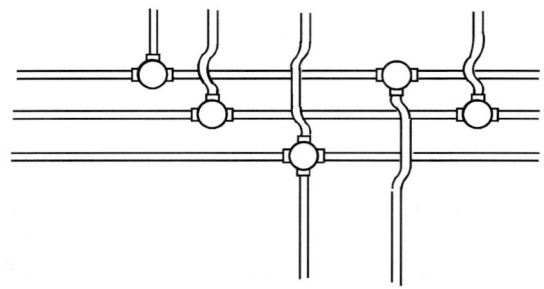

Figure 6.3

In addition to this, large-section conductors cannot easily be installed in conduits and the number of draw in points required would be considerable.

Trunking will provide good mechanical protection and has the advantages of

- much larger capacities are available
- large section conductors may be easily installed
- the detachable lid makes the installation of the cables quicker and easier
- conductors may be added without risk of damage to the cables already installed
- conduit may be connected to the trunking to take cables to individual machines, circuits and so on

So we use trunking when these situations arise, providing a much neater installation (Figure 6.4).

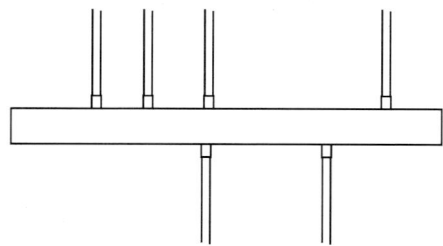

Figure 6.4

It is also common practice to use trunking where a number of conduits come together and need to take different routes, at a main distribution point for example (Figure 6.5).

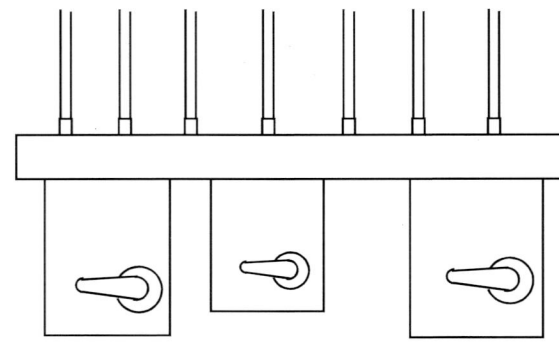

Figure 6.5

This provides a neat and more adaptable system. We may use steel trunking as the circuit protective conductor for the circuits installed, but we must take care to ensure good electrical and mechanical continuity throughout, regardless of whether the trunking is used as a cpc. To see how this is achieved we must consider how a trunking system is put together.

Assembly

Steel trunking can be easily cut to length with a hacksaw, and care should be taken to make sure that cut ends are square and all burrs are removed. Some form of packing should be used to prevent distortion when trunking is held in a vice, and a block of wood fitted inside the trunking, as shown in Figure 6.6, will normally suffice.

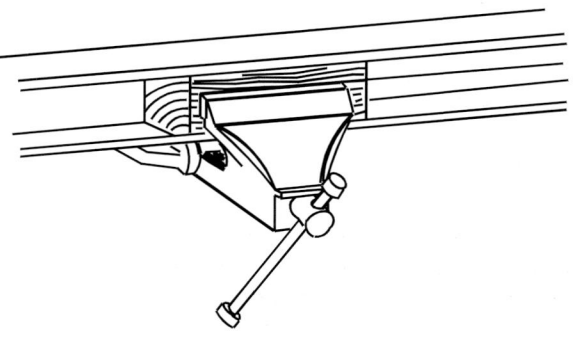

Figure 6.6

Our lengths of trunking are joined together, in most cases, by the use of couplers. The type of coupler used (Figures 6.7 and 6.8) will depend on the manufacture of the trunking, so it is advisable to use the same make throughout the system.

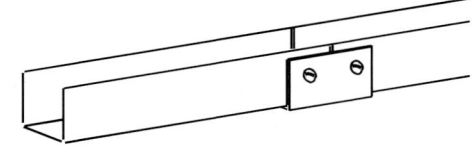

Figure 6.7   Outside coupler

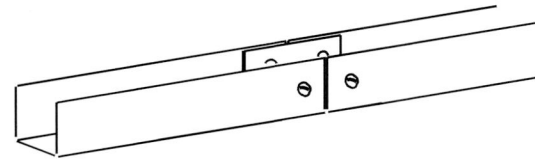

Figure 6.8   Inside coupler

Where joints are made in this way it is important to ensure earth continuity across the joint, and this is normally carried out using copper straps across the join (Figure 6.9). These may be bought purpose-made or a punched copper strip (Figure 6.10) may be cut to length.

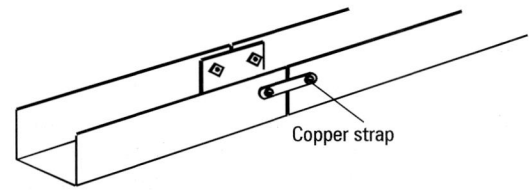

Copper strap

Figure 6.9

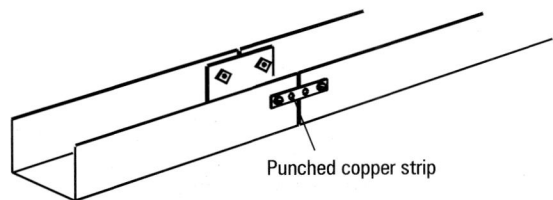

Punched copper strip

Figure 6.10

Although this can now be used as a circuit protective conductor it must NOT be used to provide protective earth and neutral conduction. BS 7671 states that metal enclosures for cables may not be used as Combined Protective Earth and Neutral Conductor (PEN).

Trunking bends

Changes in the direction of trunking can be achieved by either
- bends, elbows, tees and other manufactured fittings
- fabrication on site

Typical **manufactured fittings** are shown in Figure 6.11 and it is important to remember that the lid must be accessible at all points, so the correct type of fitting must be used to ensure that this is the case.

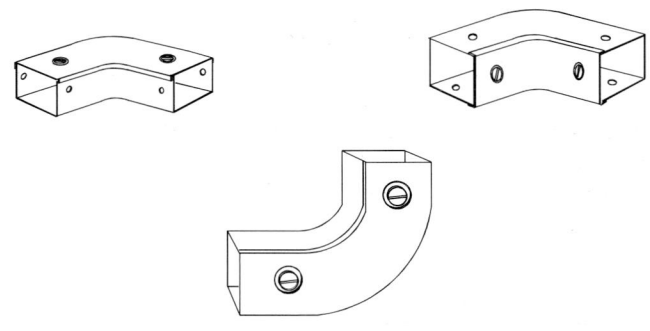

Figure 6.11   Manufactured fittings

Site-fabricated bends and sets (Figure 6.12–6.14) may be constructed by cutting and joining trunking on site. The jointing process may be done by using bolts, rivets or welding. Bolts are most commonly used. Trunking fittings should comply with BS4678, and therefore fabrication on-site is rarely undertaken. The number of cables allowed to be installed in the "tee" in Figure 6.12 would be reduced by this type of on-site manufactured tee therefore it is not widely used on trunking runs.

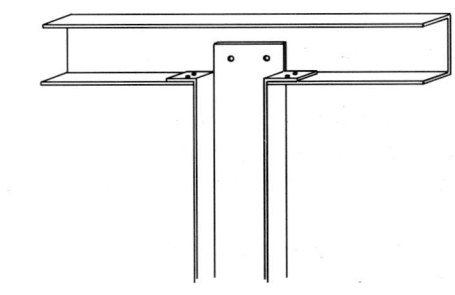

Figure 6.12   "Tee"

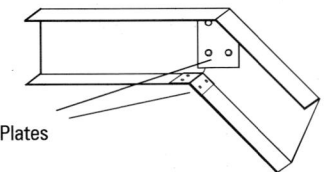

Plates

Figure 6.13   Bend

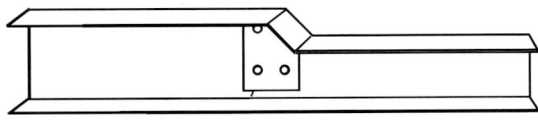

Figure 6.14   Reduction

We must ensure good mechanical and electrical joints between conduit and trunking, and trunking and switchgear. To do this we should remove the protective enamel on the inside of the trunking where these joints are made (Figure 6.15), and in the case of conduits a serrated washer should be fitted between the conduit and the trunking (Figures 6.16 and 6.17).

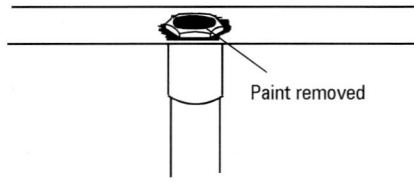

Figure 6.15   Protective enamel paint removed

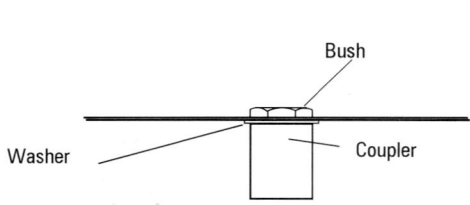

Figure 6.16   Serrated washer fited between the conduit and trunking

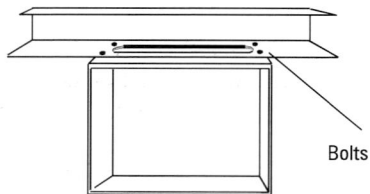

Figure 6.17   Switchgear attached by bolts to trunking

Remember

- **Metal cable enclosures may NOT be used as PEN conductors.**

- **Trunking lid must be accessible for removal at all points.**

- **Burrs and sharp edges must be removed from all trunking and lid.**

- **Joints should be mechanically and electrically sound**

- **Bonding straps should be used to ensure electrical continuity.**

There are some special requirements for metal trunking installations which we must consider during the erection of the system.

Remember that we were made aware of the electromagnetic effects of steel conduits in Chapter 4. We face similar problems with any type of steel enclosure, and metal trunking is no exception. In addition to the precautions that we had to take for conduit, we must make sure that (Figure 6.18)

- where cables pass through holes in the trunking, into switchgear for example, that all the cables associated with an a.c. circuit pass through the same hole
- that we slot metal boxes and the like to restrict the build up of eddy currents
- that we do *not* use single core SWA cables on a.c. circuits

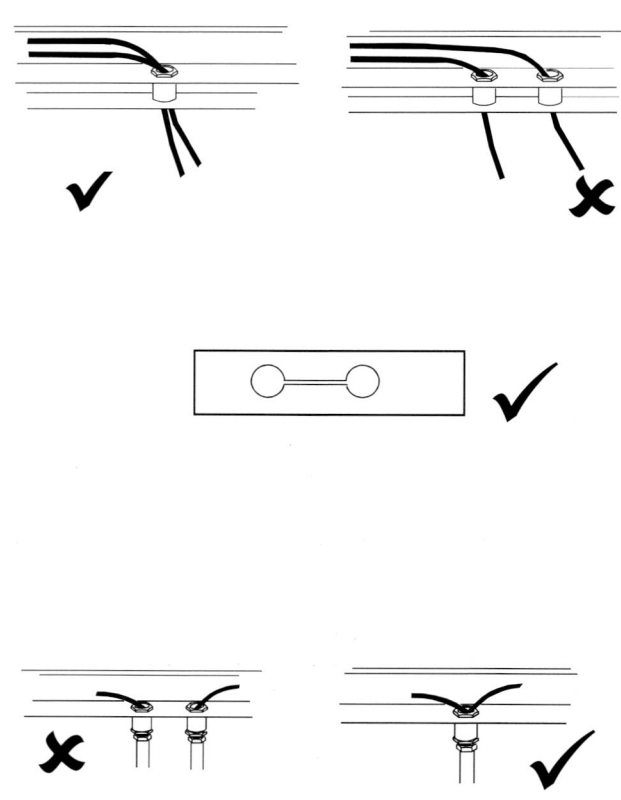

Figure 6.18

Remember

We may use a high current a.c. ohmmeter to test the continuity of a steel conduit or trunking. An Earth Fault Loop Impedance Tester may prove useful.

Where cables pass through holes in trunking these must be bushed to prevent damage to the insulation. If large holes or slots are used, at distribution boards for example, a grommet strip may be used (Figure 6.19). This is a plastic or neoprene strip with a moulded groove, rather like a rubber grommet. It is pushed onto the trunking and the interference fit ensures it remains in place. We must ensure it is cut accurately to length to give complete protection.

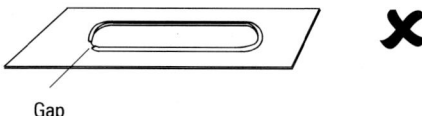

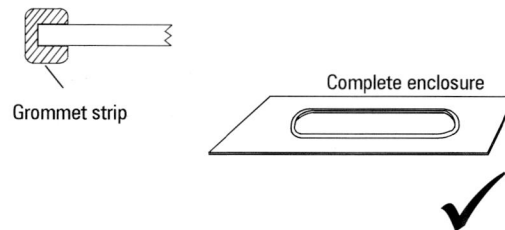

Figure 6.19

There are three other important points to consider with any trunking installation.

The first is regarding vertical runs of trunking. Remember that if we have a long vertical run of trunking we may need to fix cable supports into the trunking to support the cables enclosed (Figure 6.20).

As a general rule the maximum distances that cables may be run unsupported in vertical trunking runs is 5 metres. If our run exceeds this we must use an intermediate support.

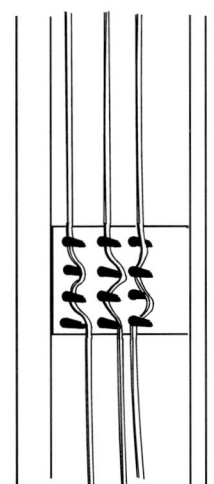

Figure 6.20   Vertical trunking run with cable supports

Secondly, our trunking can provide a means for fire and smoke to move from one area to another. In fact, as cables carrying current produce heat, a vertical run of trunking can reach high temperatures at the top end if precautions are not taken to prevent this from happening.

We do this by installing fire and heat barriers within the trunking (Figure 6.21):
- for vertical runs of trunking fire barriers must be installed where trunking passes through floor levels
- where trunking passes through a structural fire barrier, such as a brick wall, an internal fire barrier must be fitted to give at least equal protection against the spread of fire through the trunking.

These are generally of a mineral wool type of construction.

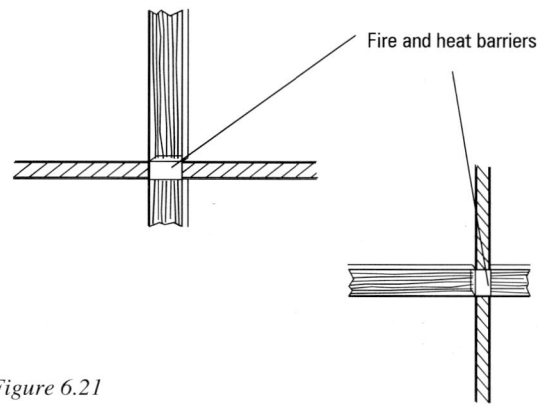

Figure 6.21

The final consideration we shall look at is the support of the trunking itself. There are tables available which show the maximum distance between supports for the trunking itself.

The most common method of fixing trunking is to drill the back of the trunking and fix it directly to the structure of the building. It may be necessary to support trunking on brackets or hangers, depending on the location and type of structure. Some typical fixings are shown in Figure 6.22.

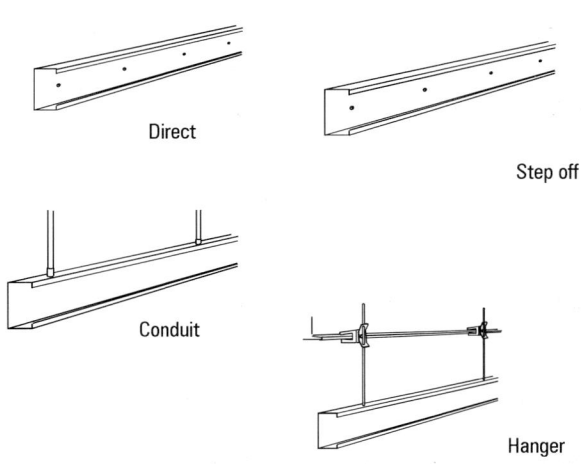

Figure 6.22   Trunking brackets and hangers

79

Try this

Using IEE Guidance Note 1 check up on the spacing of supports for the trunking sizes given:

| c.s.a. of trunking mm² | Maximum distance between supports | |
| | Metal | |
	Horizontal m	Vertical m
> 300 < 700		
> 700 < 1500		
> 1500 < 2500		

Points to remember ◄ – – – – – – – – – – – – –

Steel trunking is made of low carbon steel and supplied in _____ metre lengths having sections between 38 mm × 33 mm and 225 mm × 100 mm.

Steel trunking may be used as a cpc but it cannot be used as a PEN conductor. Joints must be _____ across to provide continuity. Fire barriers must be fitted wherever trunking passes through a natural fire barrier and vertical runs must have fire barriers at each floor.

The most common method of fixing trunking is to _____ the back of the trunking and fix it directly to the structure of the building.

What may be used if it is necessary to support trunking?

Self-assessment multi-choice questions
Circle the correct answers in the grid below.

1. General purpose trunking is supplied in lengths of
   (a) 2.0 metres
   (b) 2.5 metres
   (c) 3.0 metres
   (d) 3.5 metres

2. (i) Copper straps are used to provide continuity across trunking joints.
   (ii) Trunking can be used as a PEN conductor.
   (a) Both statements (i) and (ii) are correct.
   (b) Statement (i) only is correct.
   (c) Statement (ii) only is correct.
   (d) Both statements (i) and (ii) are incorrect.

3. The maximum distance between cable supports in vertical runs of trunking is generally accepted as
   (a) 9 metres
   (b) 7 metres
   (c) 5 metres
   (d) 3 metres

4. If trunking passes through a brick wall from one room to another we must fit
   (a) a coupling
   (b) a fixed lid section
   (c) a grommet strip
   (d) a fire barrier

5. A steel trunking with a c.s.a. of 500 mm² run horizontally must be fixed at a maximum interval of
   (a) 0.75 metres
   (b) 1.25 metres
   (c) 1.75 metres
   (d) 3.0 metres

Answer grid

1	a	b	c	d
2	a	b	c	d
3	a	b	c	d
4	a	b	c	d
5	a	b	c	d

Part 2

Specialist trunking

In addition to our standard trunking there are a number of special trunkings constructed for a particular purpose.

Compartmentalised trunking

Whilst this is a special it is also available in the form of standard, skirting and floor trunking. "Compartmental" implies some form of separation or barrier. This type of trunking separates circuits that are in different voltage bands.

There are in fact two voltage bands, these being

Band I: this includes installations where
(a) protection against electric shock is provided by the value of voltage
(b) the value of voltage is limited for operational reasons, such as telecommunications and computers and alarm circuits.

Band II: this includes the voltages for the supply to most commercial, industrial and domestic installations. LV (Low voltage) will usually be within voltage Band II which does not include voltages exceeding 1000 V a.c. or 1500 V d.c.

If Band I and Band II voltage circuits are to be installed in a common wiring system, BS 7671 details a number of conditions, one or more of which must be achieved in each case. Many of these relate to the insulation and construction of the cables used which, for practical reasons, cannot always be achieved. For example we may have to install bell call circuits, telephones and so on using conductors insulated up to 600 or 1000 volts, making them too big to fit into the typical accessories and so making the installation very costly. Some computer and data transfer circuits cannot be run with Band II circuits without expensive screening again causing an increase in size and cost.

However, one option available is that Band I and Band II circuits are contained in separate conduit, trunking or ducting systems. To achieve this requirement it is common practice to install a compartmentalised trunking system (Figure 6.23). The compartments then keep the different voltage Bands completely segregated (separate) throughout.

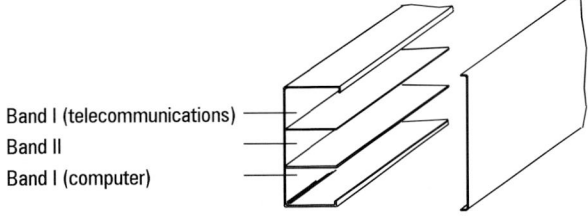

Band I (telecommunications)
Band II
Band I (computer)

Figure 6.23   Compartmentalised trunking system

Using this type of system we can easily comply with all the requirements for different voltage bands and systems.

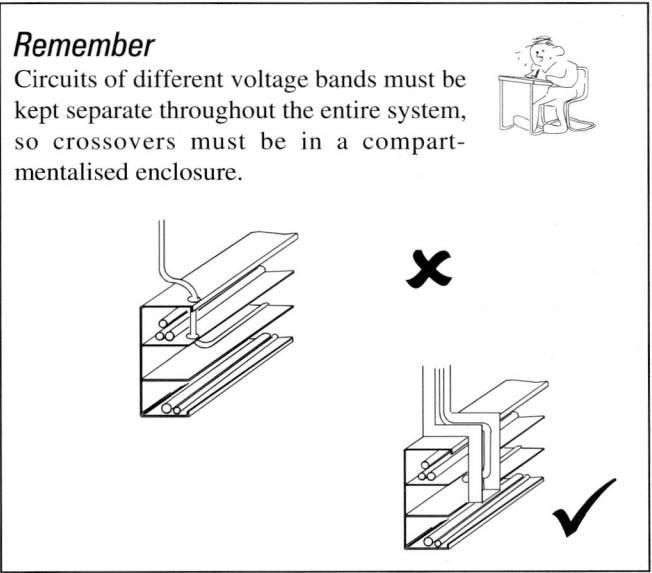

Remember
Circuits of different voltage bands must be kept separate throughout the entire system, so crossovers must be in a compartmentalised enclosure.

Skirting trunking

Skirting trunking, as the name implies, is run at skirting level around the room.

If it is necessary an architrave trunking (Figure 6.24) may be installed to get around doorways and the like.

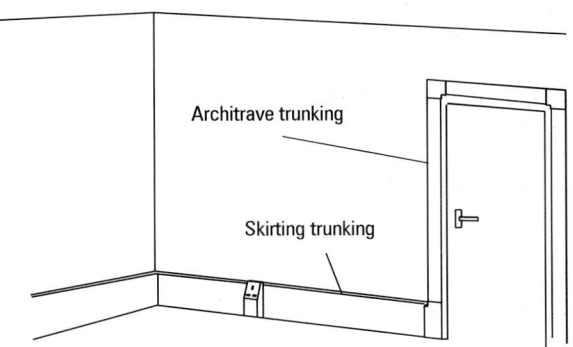

Architrave trunking

Skirting trunking

Figure 6.24   Skirting and architrave trunking

This trunking is usually much wider than a normal skirting board and has the following advantages:
- it provides good mechanical protection
- additional circuits and outlets can easily be added with minimum disruption
- socket outlets and so on can be fitted directly into the trunking
- compartmentalised trunking may be used so Band I voltage circuits can be segregated from the Band II circuits

These advantages result in it often being used in office and public building installations where change of use or equipment demands an unobtrusive and easily adaptable system offering good mechanical protection.

Try this
Use a manufacturer's catalogue to list the trunking, fittings and end caps required to install a skirting trunking around the room shown.

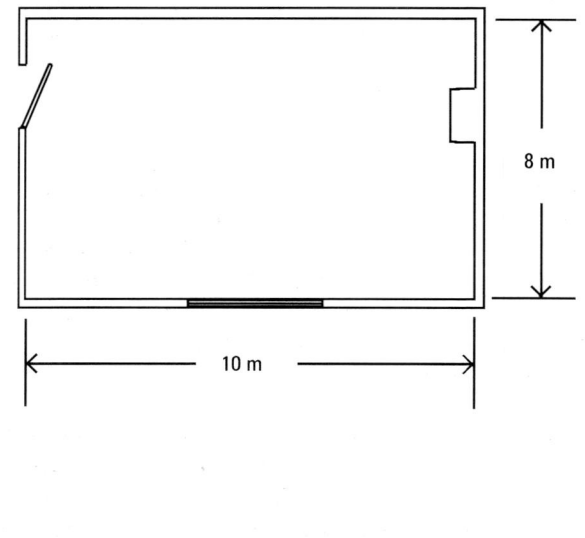

8 m

10 m

Dado and bench trunking

Both these types of trunking, like the skirting trunking, are wide enough to have the accessories mounted directly onto the trunking.

Bench trunking is used, as the name suggests, around benches in laboratories and so on. It is usually of a wedge construction allowing for easy access to outlets and a good clearance for flex bending radii.

Dado trunking (Figure 6.25) is often used around open plan offices, sometimes combined with a floor trunking system. It is generally mounted above desk height so as not to interfere with the positioning of office equipment. In order to get cables around doorways, either a floor trunking link or architrave trunking is used.

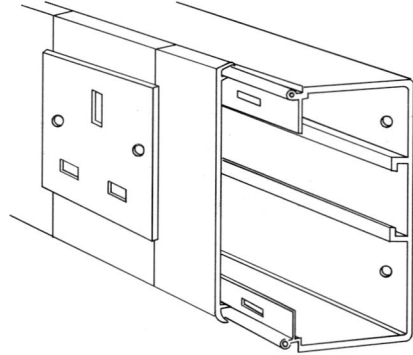

Figure 6.25   Dado trunking

PVC trunking

The types of steel trunking mentioned so far are also available in High-Impact PVC. Probably the most commonly used PVC trunking is Dado and mini trunking. PVC trunking is ideally suited for domestic installations and electronic equipment. Use of large PVC trunking is less common, although in some locations it does provide an ideal solution. For example, it is sometimes used in industrial situations because of its excellent properties of electrical insulation and resistance to corrosion.

The typical sizes of PVC trunking range from 20 mm × 15 mm to 100 mm × 50 mm, but it is available in larger sizes, usually to order. Standard accessories such as elbows, bends, tees and flanges are readily available and may be used to construct an installation in the same way as we do with our steel accessories.

PVC trunking is easily worked, and providing we observe the minimum cable bending radius requirements, can easily be cut and formed on-site. Cutting on site with a saw and mitre block enables us to produce bends and tees using solvent adhesive and plastic reinforcing sheet.

The connecting sleeves and fittings are designed for an interference fit and may not require adhesive. Should we need to apply an adhesive this should be applied to one end of the fitting only, usually the base. A gap of approximately 1 mm per metre length should be left to allow for expansion and contraction.

If there is a possibility of moisture ingress then a non-setting mastic sealer should be applied as appropriate. This should be done with care so as to maintain the neat appearance of the installation.

Mini-trunking

This is a plastic trunking with a snap-on lid manufactured in white PVC with sections from 16 mm × 16 mm to 50 mm × 25 mm as a single compartment and 38 mm × 16 mm to 38 mm × 25 mm for twin segregated trunking (Figure 6.26). Both types are supplied in standard 3 metre lengths.

In real terms, mini-trunking does supply extra mechanical protection for PVC/PVC cables, and its biggest advantage is that it provides a very neat installation. It is often used to provide surface wiring where chasing in is not practical and mini-trunking provides an ideal alternative. The use of twin compartments means that we can also provide separation for different the voltage bands.

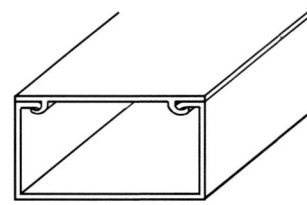

Figure 6.26   Mini-trunking

Because of its small physical size it is not very practical to manufacture bends and sets in trunking on-site, so manufactured bends, tees and intersections are normally used (Figure 6.27).

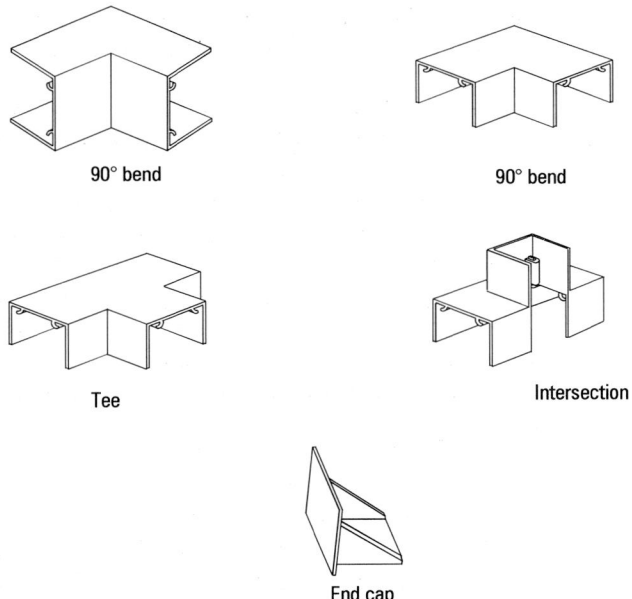

90° bend

90° bend

Tee

Intersection

End cap

Figure 6.27

You will notice that these fittings take the form of lid sections. This is because unlike our trunking system this trunking is providing extra mechanical protection and is not a completely sealed system. The method of installation may require trunking to simply butt up together, and the fitting completes the enclosure (Figure 6.28).

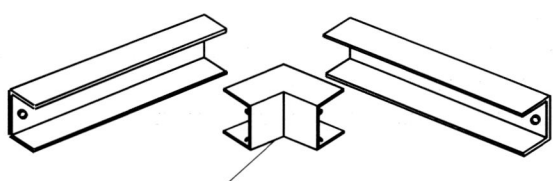

The fitting completes the enclosure

Figure 6.28

This form of construction means that a large number of different fittings are available to provide for the permutations of runs of trunking. Care must therefore be taken when we design this type of system to ensure that the correct fittings are obtained. It is not practical to show all the varieties here, so it is best to refer to a manufacturer's catalogue to confirm the fittings that you will need.

To complete the system we must be able to terminate the trunking into outlet boxes for socket outlets, switches and so on. We have two alternatives, the first of which is to use special boxes which have a form of spouted entry into which the trunking may be fitted (Figure 6.29).

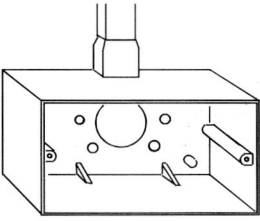

Figure 6.29

Our other alternative is to use an adaptor that fits onto the box and accepts the trunking (Figure 6.30).

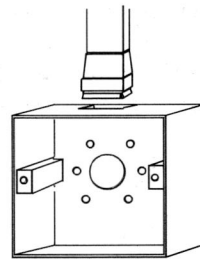

Figure 6.30

In either event, the type of box used is not usually the hard plastic moulded box we use on PVC surface wiring systems. It is usual to use a more malleable uPVC box with larger rectangular entries to accommodate the trunking.

Remember
If more than one cable is enclosed in a mini-trunking then the cables must be rated as being bunched.

This system may be used to form an entire system or to provide protection for switch drops and sockets from ceiling level (Figures 6.31 and 6.32).

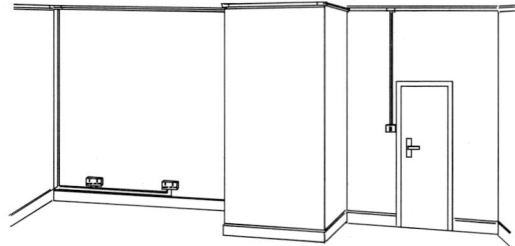

Figure 6.31

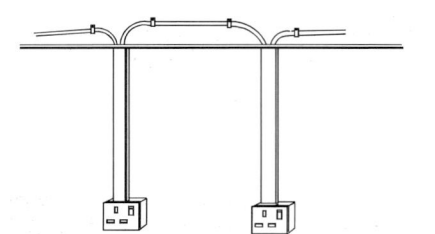

Figure 6.32

Lighting trunking

Lighting trunking is specially designed to support lighting fittings attached directly to the trunking.

To make this possible the trunking has specially shaped lips that both strengthen the trunking structure and provide fixings, via special fittings, to which the lights may be attached. Both construction and fixings are shown in Figures 6.33 and 6.34. The lid to this trunking faces downwards and is usually of the clip-in type. It may be necessary to install support clips for the cables if the light fittings are spaced far apart, as the cable will tend to hang in loops until fittings and lid are installed.

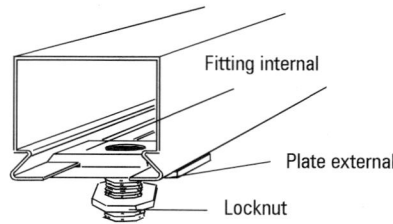

Figure 6.33

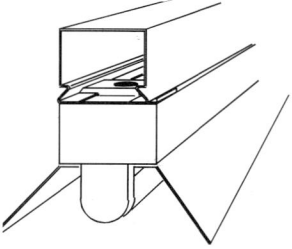

Figure 6.34   Lighting trunking and fitting

Floor trunking

Floor trunking is laid in the floor, usually before the final cement screed is laid, and is positioned so as the top edge of the trunking is flush with the finished floor level. It is designed to take and support a heavy duty lid again flush with the floor. Some manufacturers allow a lipped lid so that the floor covering may be cut to fit inside the lid section, giving the finish shown in Figure 6.35.

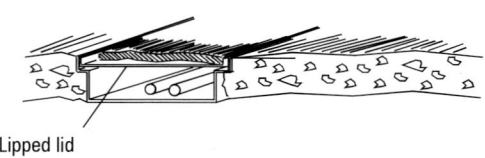

Figure 6.35   Floor trunking

Socket outlets and fittings may be fitted direct to the trunking using fitting plates (Figure 6.36).

Figure 6.36   Plate

This type of installation is often used in offices because of its versatility and it is of particular value in open plan situations where very little wall space is available. It lends itself to easy alteration due to change in use or layout and minimises the number of trailing flexes that have to be used to connect equipment. If compartmentalised trunking is used we can also house voltage Band I cables in this same enclosure.

Busbar trunking

The last special trunking we shall consider is busbar trunking. This is unlike any of those we have seen up to now, in that cables are not installed into it. Instead it contains copper or aluminium busbars which are jointed so as to be continuous throughout the length of the trunking system. At regular intervals along each length access covers are provided, and we may tap into the trunking, using tap in boxes, at any of these points.

The busbar trunking is supplied via fuses with large current ratings or circuit breakers and are generally 3 phase and neutral. The ratings of the busbar trunking and the protective devices supplying it are selected according to the total load that is to be connected, and this could be in excess of 300 A.

Each tap in, or plug in box as they are sometimes called, contains fuses or circuit breakers. In this way each individual machine, or circuit connected into the busbar, will be individually protected at its own tap in point.

This gives us a distribution system throughout a factory which has the following advantages:

- great flexibility
- easy adaptation to suit a new layout or equipment
- easy relocation of equipment without the need for major rewiring
- the ability to isolate individual machines without stopping production
- tap in points usually close to equipment, minimising circuit length, cost and so on
- small voltage drop over long runs
- emergency stop equipment can be installed to operate the whole busbar system, so shutting off all machinery with a single operation

These advantages make this system (Figure 6.37) very popular in many engineering works, factories and so on, and whilst the initial installation is quite expensive it can produce large savings in the long run.

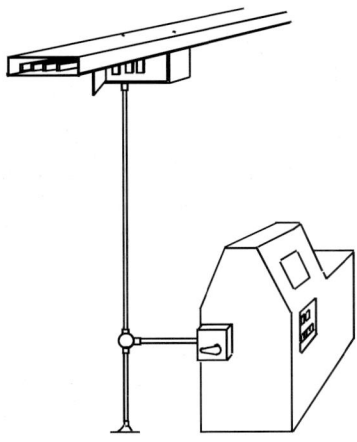

Figure 6.37   Busbar trunking system

A popular variation of this busbar trunking is used for socket outlets in offices where a raised floor is installed. The busbar trunking is installed below the raised floor and has outlets at regular intervals along its length. Socket outlets for use in the office are fitted in "floor boxes", as illustrated in Figure 6.38.

These floor boxes are cut into the raised floor tiles and connected to the busbar trunking via a trailing lead, usually in a fibre flexible conduit, and a special plug to make the final connection (Figure 6.39). This system is particularly popular because of the flexibility it offers for positioning and moving of socket outlets to suit differing office layouts.

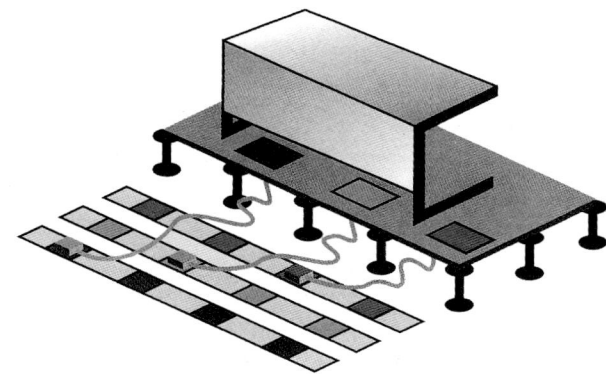

Figure 6.38

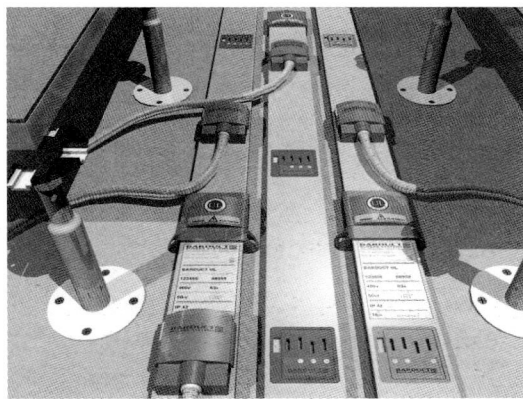

Figure 6.39

Points to remember ◄ — — — — — — — — — — — — — —

Special trunkings are also used including

skirting trunking

lighting trunking

floor trunking

_____ trunking

Compartmentalised trunking should be used where different _____ ___ _____ are to use the same enclosure.

Remember
Each circuit in PVC trunking must be installed with a circuit protective conductor.

Try this

Select the type of trunking system that you think would be most appropriate for the situations listed below.

1. To supply long runs of fluorescent lights in a warehouse.

2. To supply socket outlets in an open plan office.

3. To supply power and data cables to a computer room.

4. To accommodate large numbers of conductors at a distribution point.

5. To supply machinery in a large engineering works.

6. To supply socket outlets in an office block.

Self-assessment multi-choice questions

Circle the correct answers in the grid below.

1. If we are to install voltage Band I and Band II circuits in a single trunking the type that should be used is
   (a) skirting trunking
   (b) general purpose trunking
   (c) busbar trunking
   (d) compartmentalised trunking

2. The best type of trunking to use to supply socket outlets in a large open plan office would be
   (a) skirting trunking
   (b) floor trunking
   (c) general-purpose trunking
   (d) compartmentalised trunking

3. A recessed lid fitted to floor trunking provides a facility for
   (a) easy identification of the trunking route
   (b) inserting floor covering material
   (c) installation of socket outlets
   (d) inserting warning notices

4. When jointing lengths of PVC trunking allowance must be made for
   (a) additions
   (b) accessories
   (c) expansion
   (d) alterations

5. Where Band I voltage circuits cross Band II voltage circuits sections of a compartmental trunking we need to use
   (a) a grommet strip
   (b) bushes
   (c) crossovers
   (d) apply insulation tape

Answer grid

1	a	b	c	d
2	a	b	c	d
3	a	b	c	d
4	a	b	c	d
5	a	b	c	d

Part 3

Installing cables

In Chapter 3 we looked at the space factor for installing cables in enclosures and we will begin this part by reminding ourselves of some important points. This will include calculating the size of enclosure required for a number of cables so make sure that you have a calculator handy.

Figure 6.40

It is a good idea to look at the points we must consider when working out the capacity of an enclosure before we go through the working out involved.

- when we install cables we must ensure that no damage is done to the conductor, its insulation or the enclosure itself.
- there must be sufficient space for air to circulate around the cables once they have been installed.

> ### Remember
> The current-carrying capacity of a cable will be affected if it is grouped or bunched with other cables.

When we work out the capacity of an enclosure we do not make any allowance for the grouping of cables. The cable size must first be calculated taking this bunching into account, and the capacity is then determined based on the sizes selected.

Now there are two methods that we may use to work out the capacities of enclosures, and we shall consider each of these in turn.

Space factor

This factor is concerned with the amount of space available within an enclosure that may be filled with cables to ensure we comply with the points listed earlier. Generally a maximum value for the space factor of 45% is used. This means that only 45% of the available space may be cables, so 55% must be air space, as shown in Figure 6.41.

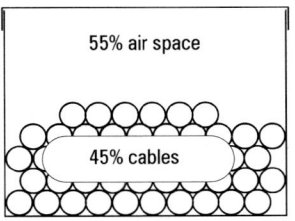

Figure 6.41

We can best illustrate the use of the space factor by using an example.

Remember that the space occupied by cables includes the insulation, so we cannot use the conductor size (such as 2.5 mm^2) as this does not include the insulation. We must therefore measure the diameter of the cable, including the insulation, and calculate its total area. Note that the effective cross-sectional area of a non-circular cable is taken as that of a circle of diameter equal to the major axis of the cable.

Suppose we need to find out how many cables of overall diameter 3 mm we can install in a 50 mm × 50 mm steel trunking.

Total area of trunking

$$50 \times 50 = 2500 \text{ mm}^2$$

Space available for cables (45%)

$$2500 \times \frac{45}{100} = 1125 \text{ mm}^2$$

This leaves 2500–1125 = 1375 mm^2 of air space within the trunking.

Space occupied by one cable

$$\frac{\pi d^2}{4} = \frac{\pi \times 3^2}{4} = 7.07 \text{ mm}^2$$

The maximum number of these cables we can install

$$= \frac{\text{space available for cables}}{\text{space occupied by 1 cable}}$$

$$= \frac{1125}{7.07}$$

$$= 159.1$$

The maximum number of these cables we could install in a 50 mm × 50 mm trunking is **159**.

Remember that we always round down to the nearest whole number, never round up.

If we needed to install conductors of different sizes we would have to do a lot of measuring and calculating to work out the capacities of our enclosures. We would usually only use this method for non-standard sizes of cable or enclosure. For standard sizes we can use the second option for calculating capacities, the use of factor tables.

Tables of factors

These are given in IEE Guidance Note 1 (Selection and Erection) and can save a great deal of time and calculation. They are based on the standard sizes of cables and enclosures.

Making reference to IEE Guidance Note 1 we find that, for example, the factor given for a solid 1.5 mm^2 PVC conductor is 8.0 while that for a stranded 1.5 mm^2 PVC conductor is 8.6. This relates to the overall diameter of a stranded conductor being larger than a solid conductor of the same cross-sectional area (Figure 6.42). These factors are used to determine trunking capacity and in themselves have no units.

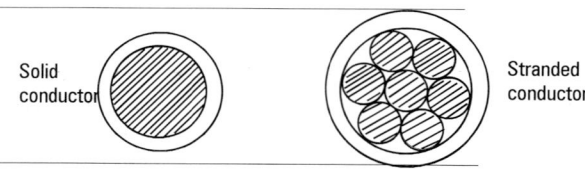

Figure 6.42

IEE Guidance Note 1 also gives factors for the size of trunking to be used. For example, a 50 mm × 50 mm trunking has a factor of 1037 whilst the factor for a 75 mm × 25 mm trunking is 738. Again these factors are relative to the area available for cable and have no units.

We must take care to use the correct factor for the type of cable that is to be installed. Once again we shall work through an example to see how the tables are used. We will use our 50 mm × 50 mm trunking and find out how many 1.5 mm^2 stranded conductors we can install in it.

Factor for 50 mm × 50 mm trunking = 1037

Factor for 1 × 1.5 mm^2 stranded cable = 8.6

Maximum number of cables we could install

$$\frac{\text{Trunking factor}}{\text{Factor for cable}} = \frac{1037}{8.6} = 120$$

Maximum number of 1.5 mm^2 stranded cables we can install in a 50 mm × 50 mm trunking is 120.

This is a much quicker method of working out capacities of trunking but we can only use it for standard sizes. We shall now apply this method to a typical installation problem.

Example

A 75 mm × 25 mm trunking is installed and contains 26 × 2.5 mm^2 stranded cables and 20 × 4 mm^2 cables. We are to install some extra circuits which will total 12 × 1.5 mm^2 solid copper cables. Does the trunking have enough capacity for these extra cables?

From IEE Guidance Note 1 the factors we require are

75 mm × 25 mm	=	738
2.5 mm^2 stranded cable	=	12.6
4 mm^2 stranded cable	=	16.6
1.5 mm^2 stranded cable	=	8.6

Total factor for 2.5 mm^2 stranded cables
26 × factor = 26 × 12.6 = 327.6

Total factor for 4 mm^2 cables
20 × factor = 20 × 16.6 = 332

Total factor for installed cables
327.6 + 332 = 659.6

Trunking factor for 75 mm × 25 mm = 738

Factor available for extra cables:
Trunking factor – total factor for installed cables
738 − 659.6 = 78.4

Factor for 1.5 mm^2 solid copper cable = 8.0

Number of extra cables we could install
$$\frac{78.4}{8.0} = 9.8$$

so we could only install 9 extra cables

Therefore our trunking has not got sufficient capacity to allow us to install 12 more 1.5 mm^2 solid cables.

***

Try this

Using the factors given in IEE Guidance Note 1 how many 6 mm^2 cables could we install in a 75 mm × 37.5 mm trunking?

Cables in conduit

So far we have looked at capacities for trunking, so now we shall consider a conduit system. Similar methods may be used to work out the capacity of conduit. However, before we begin we need to clarify some of the factors used with the tables given in IEE Guidance Note 1.

A short straight run is between draw in points – not the total length of conduit used. Any point where cables can be drawn in or out of the system can be described as a draw in point and can have a maximum length of 3 metres (Figure 6.43)

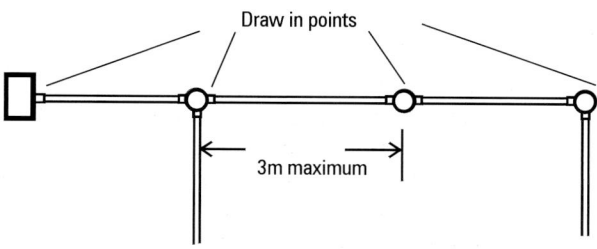

Figure 6.43

A bend is taken to be a 90° bend to the appropriate radius, and we assume that two sets of 45° are equal to one 90° bend and so on.

IEE Guidance Note 1 contains two sets of tables. One, A1 and A2, is for conduits and cables installed in short straight runs, and the other, A3 and A4, is for conduits and cables installed in runs over 3 m or containing bends and sets. First let us look at the situation for "short straight runs" using the information provided in IEE Guidance Note 1.

For short straight runs of conduit the capacity can be worked out in the same way as for trunking.

Example

We need to know how many 1.5 mm^2 cables we can install in a 20 mm conduit. We will also see what difference it would make if we use solid or stranded cables.

Conduit factor for 20 mm	= 460
Cable factor for 1.5 mm^2 solid	= 27

Maximum number of solid cables:

$$\frac{\text{conduit factor}}{\text{cable factor}} = \frac{460}{27} \quad \textbf{= 17 cables}$$

Cable factor for 1.5 mm^2 stranded = 31

Maximum number of stranded cables $\dfrac{460}{31} = 14.8$

$$\textbf{= 14 cables}$$

So we can see that if we use solid conductors we can install three more cables than we could if the conductors were stranded.

When we install conduit it is inevitable that there will be locations where we have bends or sets. The tables of factors for bends and long straight runs are given in tables A3 and A4 in IEE Guidance Note 1.

If we refer to the table A4 for runs incorporating bends and long straight runs we find that the table for conduit factors is different from those we have used so far. It is arranged in a grid with length of run down the left-hand side and the number of bends across the top.

Remember that the length of run is between draw in points. We can divide up a conduit run into shorter lengths with fewer bends by installing more draw in boxes. It is also worth considering that the use of an angle box can provide another draw in point and remove a 90° bend (Figures 6.44 and 6.45).

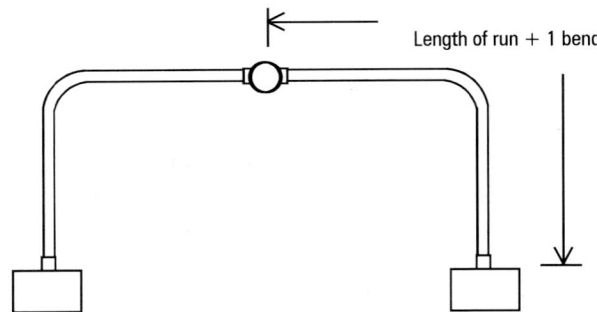

Figure 6.44

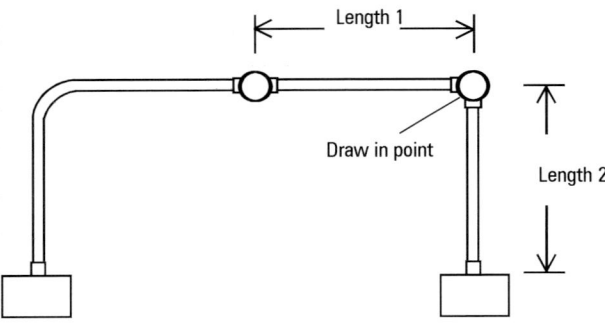

Figure 6.45

The examples that we shall consider will have the length of run stated between draw in points, so no further draw in points will be added.

We will start by looking at a 2 metre run of 20 mm conduit with one 90° bend and see how many 1.5 mm² cables we can install in it. You will notice that the table for cable factors in this case does not distinguish between solid and stranded conductors.

Conduit factor – from table A4. We find this by looking down the left-hand column until we find the 2.5 m length of run. Follow this across to the block of vertical columns for one bend and select the column for 20 mm conduit. Where these two columns cross we have a factor of 278.

90

The cable factor is selected from IEE Guidance Note 1, table A3, and is 22 for a 1.5 mm² cable.

Maximum number of cables we can install is

$$\frac{\text{conduit factor}}{\text{cable factor}} = \frac{278}{22} = 12.6$$

So 12 cables are the maximum we can install.

Now we shall use these tables to solve a typical installation problem (Figure 6.46).

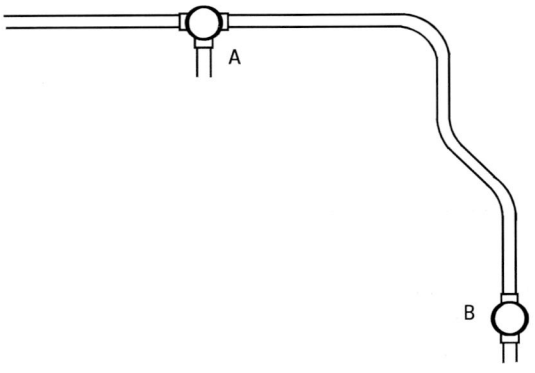

Figure 6.46

The conduit between A and B is to contain

$$6 \times 1.0 \text{ mm}^2 \text{ cables}$$

and $6 \times 2.5 \text{ mm}^2$ cables.

If the length of this conduit is 2.5 metres what is the minimum size that we can use?

Factors for cables (from IEE Guidance Note 1)

$6 \times 1.0 \text{ mm}^2$	6×16	$= 96$
$6 \times 2.5 \text{ mm}^2$	6×30	$= 180$
Total factor	$96 + 180$	$= 276$

Length of run $= 2.5$ metres

Number of bends: $1 \times 90°$ and $2 \times 45°$ sets so it is equivalent to $2 \times 90°$ bends.

Using the table from IEE Guidance Note 1 at length 2.5 m and the columns for two bends we must find a factor of 276 or more.

20 mm has a factor of 244 and so is too small.

25 mm has a factor of 442 and this is the one we must use.

So 25 mm is the smallest conduit that could be installed.

Inspection of the tables for conduit factors shows us that as the length of run or the number of bends increases the factor gets smaller, so the more complex the run the fewer cables we can install. The blank squares in the bottom right-hand corner of the table are because using that length of run and number of bends it is impractical to pull cables through. In such a case we would need to split the run up by adding more draw in points. An example of how this is done is shown in Figures 6.47 and 6.48.

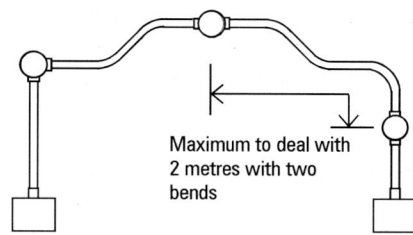

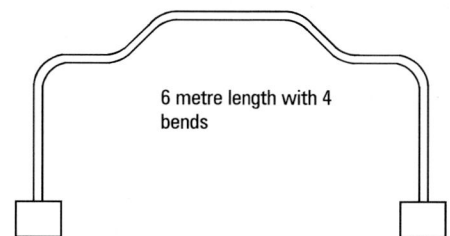

Figure 6.47

Figure 6.48

So in practical situations planning the layout with a little care can reduce cost, time and size and make the installation easier.

Try this

How many 2.5 mm^2 cables can we install in a 25 mm conduit which is 2 metres long and has two bends?

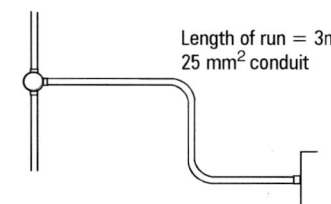

Try this

The conduit run shown in Figure 6.49 contains 8×2.5 mm^2 cables and you have to install a further 8×2.5 mm^2 cables to supply extra circuits. Is there enough capacity to allow you to do this?

Figure 6.49

Points to remember ◀ – – – – – – – – – – – – – –

The capacity of a conduit or trunking may be calculated by two methods, space factor or tables of _____
_____

Space factor is usually only applied to non-standard cables or enclosures. The space factor means that a maximum of ____% of the available space in the enclosure can be occupied by cables.

Tables of factors are given for three main types of installation
- trunking
- short straight runs of conduit (maximum 3 metres)
- long runs or runs including bends and sets

The relationship $\dfrac{\text{enclosure factor}}{\text{cable factor}}$ gives the _____ of a particular enclosure.

The length of run is the distance between _____ points.

The length of run can be reduced by adding more draw in boxes.

Try this

Which method of calculating capacities of enclosures would you use for conductors of different sizes?

What is the formula for working out the space occupied by one cable?

A 75×50 mm trunking contains 25×2.5 mm^2 stranded conductors. The number of 4 mm^2 cables that could be installed in addition, using the factors method, is
Factor for 75×50 mm trunking is 1555
Factor for 2.5 mm^2 stranded conductor is 12.6
Factor for 4 mm^2 stranded conductor is 16.6

7

Cable Tray

You will need to have a copy of IEE Guidance Note 1 available for reference in order to complete the exercises within this chapter.

Remind yourself of the following facts from the previous chapter.

The capacity of a conduit or trunking may be calculated by either of two methods: space factor or tables of _____

The space factor means that a maximum of _____% of the available space in the enclosure can be occupied by cables.

To reduce the length of run or number of _____ within a conduit run we can add more _____

The calculation of capacity of conduit and trunking capacities is to ensure that the conductor, insulation and enclosure are not damaged during the installation process.

Additional allowances must be made for the heat produced when a number of circuits are banded together in conduit or trunking.

On completion of this chapter you should be able to:

◆ describe the types of cable tray available
◆ give examples of fixings for cable tray in particular situations
◆ recognise that any bend put in the cable tray must take into account the bending radius of the cable it contains
◆ calculate the minimum bending radius of cable tray given relevant factors
◆ recognise that a correction factor may apply to the grouping of cables
◆ complete the revision exercise at the beginning of the next chapter

Part 1

Cable tray

Cable tray is a support system for sheathed cables used in industrial/commercial installations. It provides a versatile fixing in situations where fixing direct to the structure is not possible and where obstructions such as pipework makes cable runs difficult. Cable tray is particularly useful where a number of sheathed cables share a common route and where cables are to be run overhead.

Figure 7.1

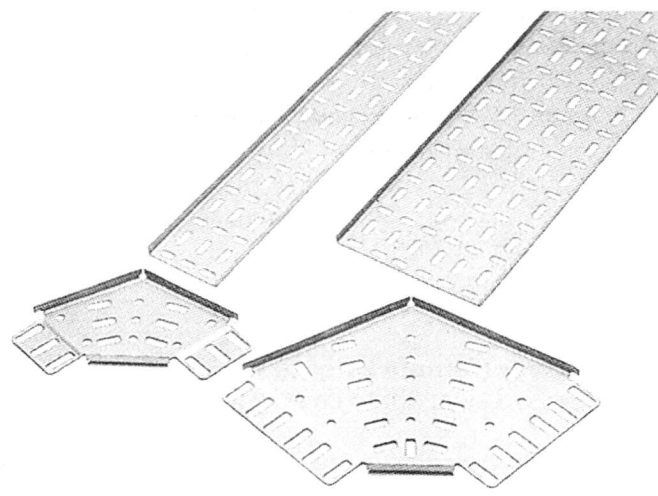

Figure 7.2

The basic construction of cable tray is a perforated steel sheet, the edges of which are folded up to form a channel. There are three main types of tray (Figure 7.3).

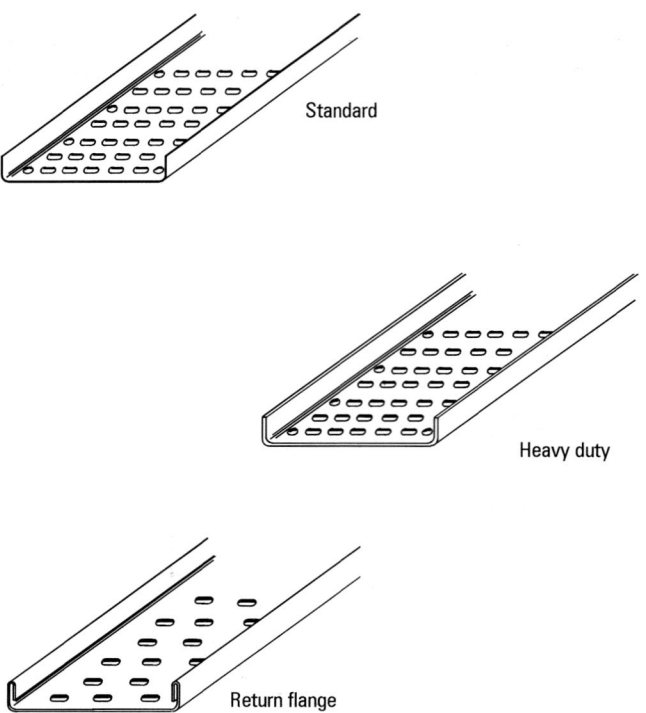

Figure 7.3

The major difference between these types is the mechanical strength of the tray itself, so the factors which govern their selection are usually

- the size and weight of the cables
- the distance between supports for the tray

Cable tray is supplied in standard lengths of 2.4 metres with widths from 50 mm to 900 mm and can be obtained with finishes to suit most environmental conditions. Some of the most common finishes are

- galvanised
- deep galvanised (thicker zinc layer)
- red oxide primed
- PVC coated
- epoxy resin coated (epoxy resin paint is also used)
- yellow chromate (fire retardant)
- stainless steel

With the exception of the stainless steel the material used for the manufacture of cable tray is usually mild steel.

Plastic and resin cable tray is also available for areas where metal tray is inappropriate.

In order to maximise the fixing potential of cable tray it has to be mounted on some form of bracket or support to allow access to both sides of tray so we can attach the cables to it. Some of the more common fixing methods are shown in Figure 7.4.

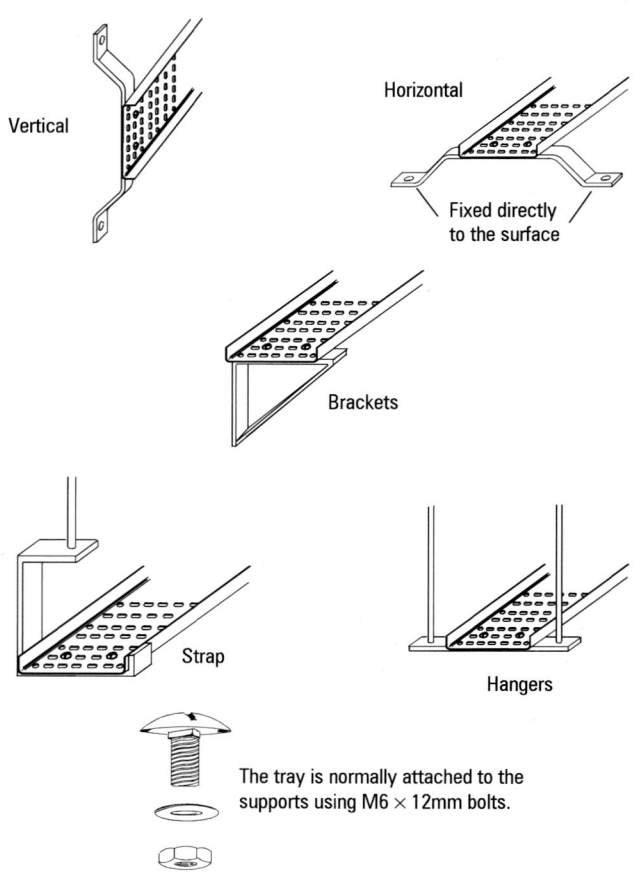

Figure 7.4

These same type of nuts and bolts are used for joining lengths of standard tray together. Return flange tray has a specially designed unique clip assembly system. This will save valuable time during installation as there is no need for further drilling where ever it is cut to length. The joining of lengths of tray is usually done by one of the methods shown in Figure 7.5.

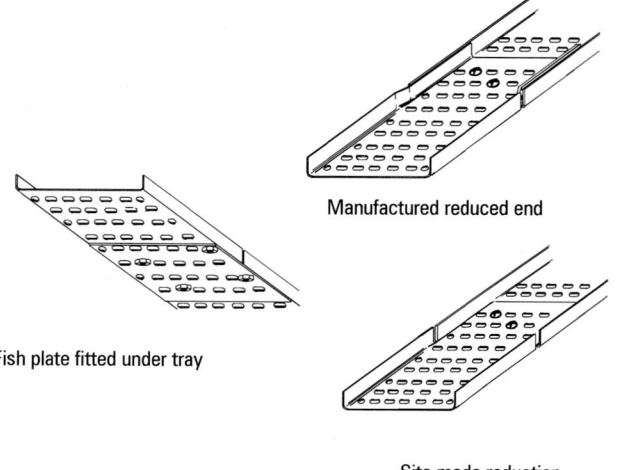

Figure 7.5

During installation changes of direction will need to be made in the route of our cable tray. This may be done by the use of manufactured fittings or by site-made angles, bends, risers and so on.

Manufacturers produce different angles for changes of direction through the same plane, such as flat bends and elbows, as well as internal and external risers and reductions (Figure 7.6).

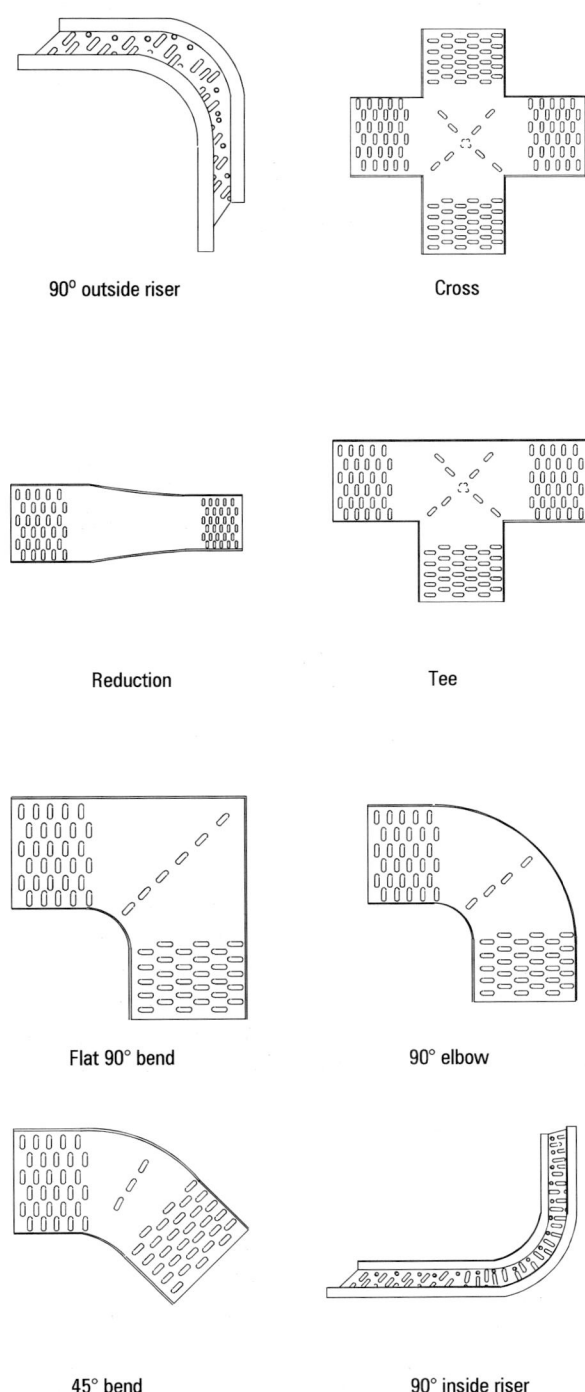

90° outside riser   Cross

Reduction   Tee

Flat 90° bend   90° elbow

45° bend   90° inside riser

Figure 7.6   Manufactured fittings

Similar effects can be achieved by cutting, jointing and bending cable tray (Figure 7.7). Cable tray can be bent using either hand tools or a proprietary bending machine.

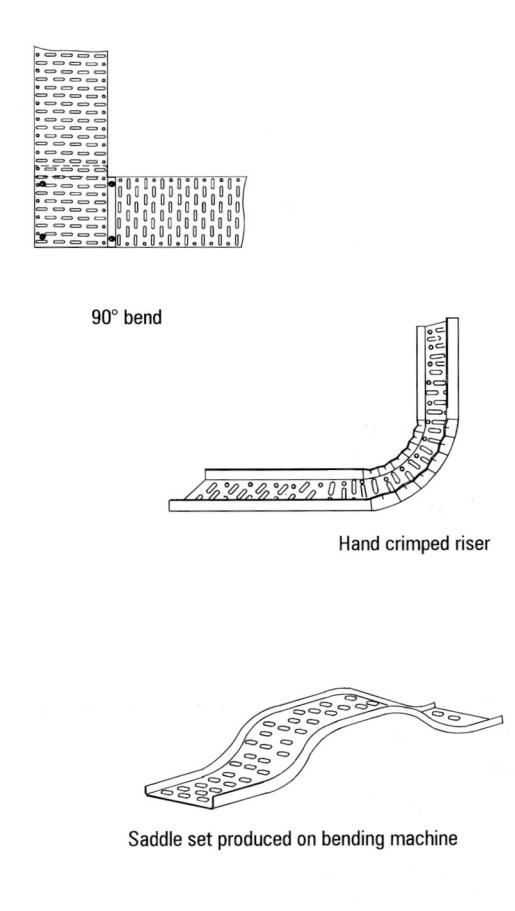

90° bend

Hand crimped riser

Saddle set produced on bending machine

Figure 7.7   Site-made fittings

There are some important points that we must bear in mind when we are installing our cable tray. Our first point to consider is the bending radius of the cables that are to be fitted to the tray. Guidance on the bending radius which must be applied to the cables is given in IEE Guidance Note 1.

If we are to install a number of different size cables on a common cable tray we must ensure that any bends or risers in the tray are *not* of smaller radius than that required by the *largest* cable installed. We will be looking at how to calculate minimum radius bends in Part 2 of this chapter.

Try this

Where have you seen any examples of cable tray installations?

Make a note of the type of building, the type of cable tray and any accessories you saw.

Alternatives to the conventional cable tray are being developed all the time. Make a note of any you have seen.

Points to remember ◄ — — — — — — — — — — — — —

The basic construction of cable tray is a perforated _____ sheet. The major difference between the three main types of tray is

Cable tray is supplied in standard lengths of _____ m, with widths from 50 mm to _____ mm.

Some of the most common finishes are

Some of the most common forms of bracket or support are

Cable tray is usually jointed by overlapping, return edge connections, bends and _____

Changes of direction in cable tray can be made by the use of manufactured fittings or by site-made angles, bends or risers. If we are to install a number of different size cables on a common cable tray we must ensure that any bends or risers are of a suitable radius for the _____ cable installed.

Self-assessment multi-choice questions

Circle the correct answers in the grid below.

1. Cable tray is supplied in standard lengths of
   (a) 3.4 metres
   (b) 3.0 metres
   (c) 2.4 metres
   (d) 2.0 metres

2. The tray fitting shown in Figure 7.8 is

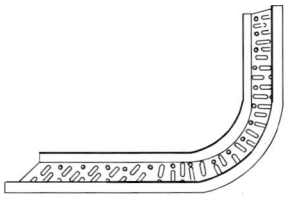

Figure 7.8

   (a) internal riser
   (b) external riser
   (c) 90° bend
   (d) 90° elbow

3. If a fire-retardant material is needed the most appropriate finish for cable tray would be
   (a) PVC coated
   (b) epoxy resin coated
   (c) yellow chromate
   (d) red oxide primer

4. The normal method for joining lengths of tray is
   (a) welding
   (b) bolting
   (c) clips
   (d) adhesive

5. The radii of bends and sets in cable tray are determined by the
   (a) thickness of cable tray
   (b) type of cable tray
   (c) number of conductors
   (d) cross-sectional area of the cables installed

Answer grid

1	a	b	c	d
2	a	b	c	d
3	a	b	c	d
4	a	b	c	d
5	a	b	c	d

Part 2

Installation of cables on cable tray

Let's look at an example.

We are to install six different sizes of SWA cable on a cable tray. The factor, from IEE Guidance Note 1, Appendix I, that must be applied to the outside diameter of PVC/SWA cable is 6.

So if our cables are of outside diameter 6 mm to 20 mm then a riser installed must have a minimum radius bend of

$$6 \times 20 \text{ mm} = 120 \text{ mm}$$

This ensures that none of the cables are bent at too tight a radius (Figure 7.9).

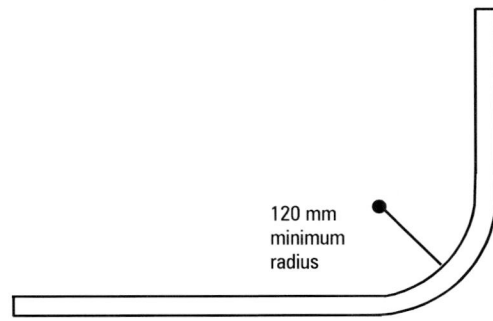

120 mm minimum radius

Figure 7.9

Example

Determine the minimum bending radius for a cable tray riser containing 4 MIMS cables of 6 mm². The factor to be applied to the outside diameter of MIMS cable is 6.

$$6 \times 6 = 36 \text{ mm}$$

The minimum bending radius 36 mm.

The second point to consider is the spacing of cables, as this will affect the width of cable tray required. We know that bunching cables together affects their current-carrying capacity, but it is worth remembering the following:
- a smaller reduction in current carrying capacity is needed if cables are "SPACED" by a distance of one cable diameter between adjacent surfaces
- if the horizontal distance between adjacent cables exceeds twice their overall diameter no correction factor need be applied (Figure 7.10)

Bearing these facts in mind and selecting accordingly can make considerable savings in cable costs.

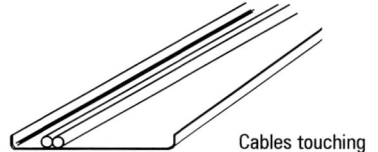

Cables touching

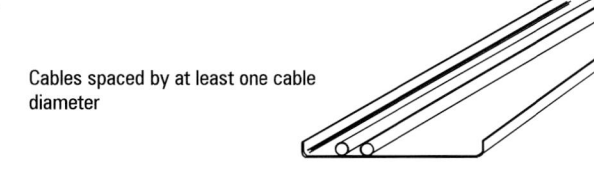

Cables spaced by at least one cable diameter

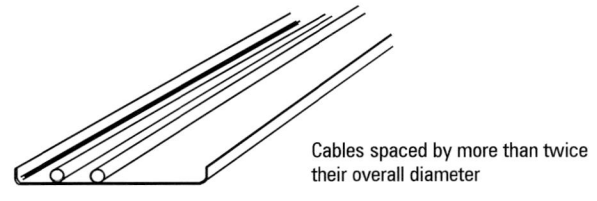

Cables spaced by more than twice their overall diameter

Figure 7.10

Try this

Given the following:
factor applied to outside diameter of MIMS cable = 6
factor applied to outside diameter of SWA cable = 6
Determine the minimum radius of bend for a cable tray carrying all of the following:

$3 \times$ MIMS cables with outside diameters of 4 mm², 6 mm², and 10 mm²

$3 \times$ SWA cables with outside diameters of 12 mm², 25 mm² and 16 mm²

Fixing cables

The final point we shall consider is the spacing between the fixings that hold the cable to the tray. If we run cable tray horizontally with the cables installed on the tray then the tray itself will provide support for the cables. If the tray is run vertically then cable support is provided by the fixings to the tray only, and we must observe the maximum distances between supports for cables. IEE Guidance Note 1 gives details, and examples should have been completed in the "Try this" on p. 26 of this book. Manufacturers' recommendations should be followed.

Having talked about fixing cables to the cable tray this is a good time to see how this can be achieved. There are a variety of ways of doing this (Figure 7.11).

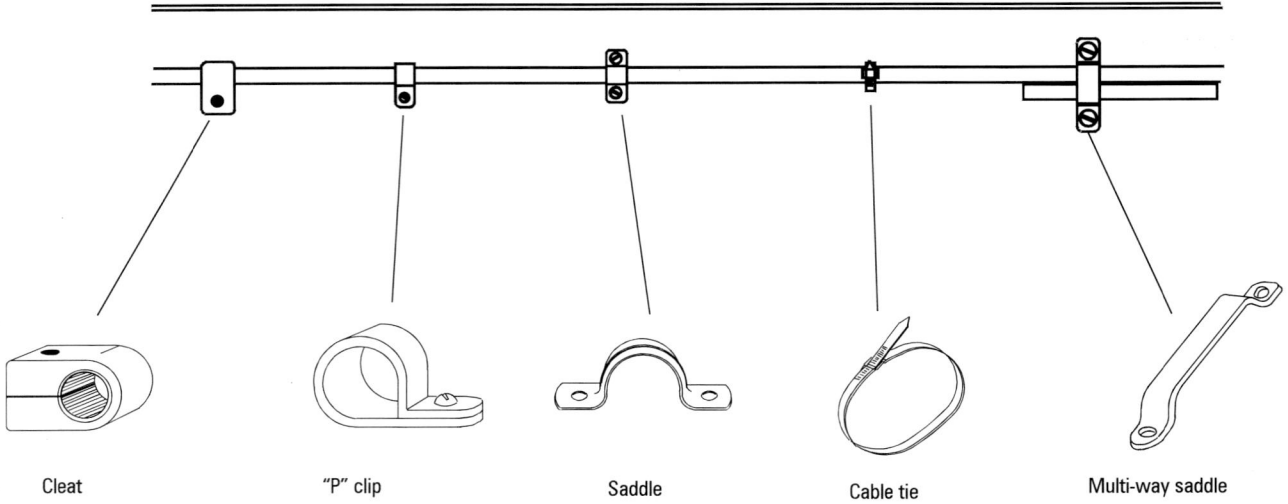

| Cleat | "P" clip | Saddle | Cable tie | Multi-way saddle |

Figure 7.11

The clip and saddle types of fixing are usually attached by screws and nuts through the perforations in the tray base. The cable tie is wrapped around the cable and the solid section of tray between the perforations in the base (Figure 7.12).

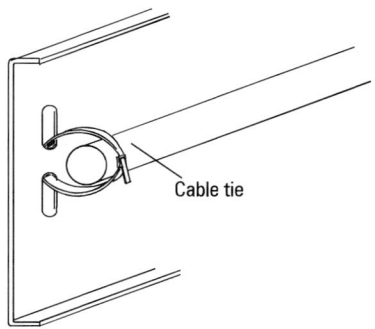

Cable tie

Figure 7.12

Sketch the most suitable method of supporting cable tray, in your opinion, in the following locations:

Run vertically up a brick wall.

Suspended below an "I" section R.S.J.

Run at 90° to a vertical surface.

Suspended below a plasterboard ceiling.

Try this

1. If 6 × 6mm outside diameter cables are to be fixed onto a cable tray without a grouping factor to be applied, the minimum width of tray would be (to nearest standard size)

2. Using manufacturers' details determine the size of cable tray required to carry the following cables if they are to be installed as "spaced".
   (a) 3 × PVC/SWA/PVC cables with overall diameter of 28 mm with 2 × PVC/SWA/PVC cables with overall diameter of 36 mm.
   (b) 4 × SWA/PVC/XLPE cables with overall diameter of 47 mm

Self-assessment multi-choice questions

Circle the correct answers in the grid below.

1. If a MIMS cable of 10 mm outside diameter is to be installed on a cable tray the minimum radius that the tray may be bent to is
   (a) 60 mm
   (b) 50 mm
   (c) 20 mm
   (d) 10 mm

2. The maximum distance between the cable clips holding the 10 mm outside diameter MIMS cable to the cable tray if it is run vertically is (Use IEE Guidance Note 1 or manufacturer's details)
   (a) 600 mm
   (b) 800 mm
   (c) 900 mm
   (d) 1200 mm

3. If no grouping factor is to be applied to cables installed on cable tray the minimum horizontal distance between the cables must be
   (a) 4 × outside diameter
   (b) 3 × outside diameter
   (c) 2 × outside diameter
   (d) 1 × outside diameter

4. The best method of fixing a number of small diameter cables to a tray if grouping is not a problem is by
   (a) "P" clips
   (b) saddles
   (c) cable cleats
   (d) cable ties

5. If a cable tray is to be run at 90° to a vertical surface the best form of support to use is

   (a)  (c)

   (b)  (d)

8

Installation Circuits

Complete the following to remind yourself of some of the important facts from the previous chapter.

Cable tray is a support system for sheathed cables in

_____ situations.

Cable tray is particularly useful where a number of sheathed cables are to be run overhead.

We are to install four different sizes of SWA cables on a cable tray. The sizes of the cables vary between 6 mm and 20 mm. If the horizontal distance between adjacent cables exceeds _____ their overall diameter no correction factor need be applied.

On completion of this chapter you should be able to:

◆ describe open and closed circuits as used for alarm systems
◆ describe the wiring of switches in a lighting circuit
◆ identify the special requirements when wiring bathroom lighting circuits
◆ identify the different types of lamps used in lighting circuits
◆ describe the special requirements of bathroom and cooker circuits
◆ describe the operating principles of the three heat switch
◆ complete the revision exercise at the beginning of the next chapter

Part 1

In this chapter we will be looking at series circuits, parallel circuits and combinations of each. These circuits are described in detail in the book "Basic Science and Electronics".

Alarm and emergency systems

Open circuit systems

The simplest circuit is that used on front doorbells, for it only contains one push, one bell and source of supply plus, of course, the cable to connect them together.

The push completes the circuit when it is pressed and is known as a "push to make" type. The "bell" may, of course, be chimes or a buzzer, but for simplicity we will continue to use the bell symbol (Figure 8.1).

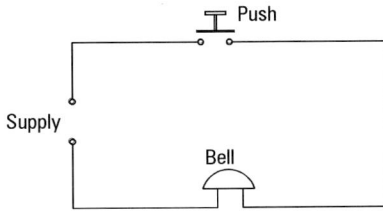

Figure 8.1   *"Push to make" circuit*

If an extra push is required in this circuit it must be connected in parallel with the first one (Figure 8.2).

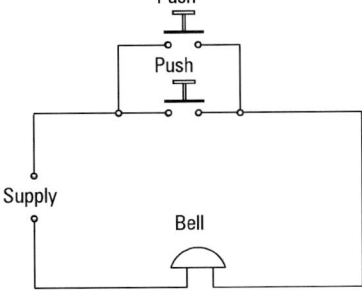

Figure 8.2

Instead of "push to make" circuits, other detection devices may be used to set off an alarm.

This type of circuit has many uses but has the problem that if a conductor is broken the whole system is out of action.

Closed circuit systems

When a bell circuit is used for emergency alarms it is more likely to be a closed circuit system (Figure 8.3). This means that the whole detection circuit is complete until something breaks it and sets off the alarm. To make the circuit practical a relay is used to divide the detection circuit from the alarm sounder circuit.

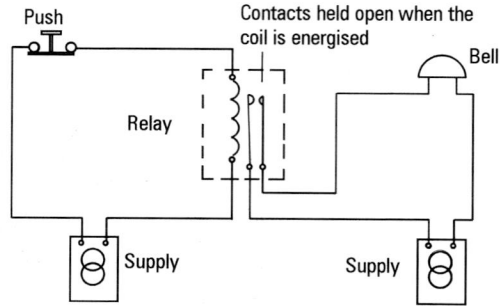

Figure 8.3

This relay is basically the switch that makes the circuit when the alarm is to be sounded. When the circuit is healthy the relay contacts are held apart by the magnetic field of the relay coil. This coil is energised all the time the detector circuit is complete. As soon as any part of the detector circuit is opened the alarm is sounded. Extra detectors in this circuit are connected in series with the first (Figure 8.4).

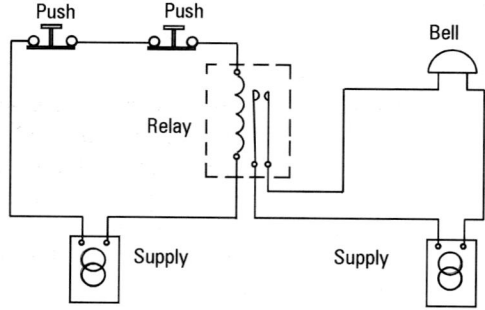

Figure 8.4

These normally closed circuits are used for intruder or fire alarms. The detectors on intruder alarms may be foil strips on windows, contacts on doors, pressure pads or many other devices. On fire alarms they may be heat detectors, smoke detectors, flame detectors or straightforward manually operated smash glass contacts (Figure 8.5).

Figure 8.5

Emergency lighting

Another emergency system we need to consider is emergency lighting. This needs to be examined from several different points of view. First, are the emergency lights to have their own emergency power supply or will this come from a central source? Next we need to determine whether the emergency lighting should be on all the time even when there is not an emergency. In this case it is referred to as being "maintained", or if it is only required to come on when the main supply fails – "non-maintained".

Central supply system

This may consist of a central battery bank (Figure 8.6) which consists of secondary cells which are constantly on charge. In a very large installation the central supply could be a standby generator. Either method would have its own distribution system and circuits wired through to where the emergency lights are required.

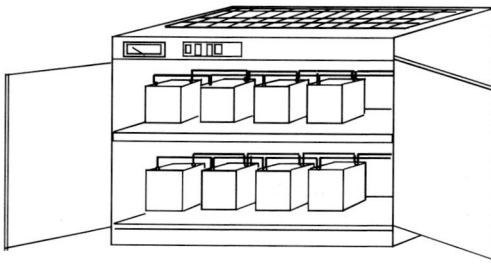

Figure 8.6   *Battery central control unit for emergency services*

Try this

Look at your front doorbell system.

How is it supplied?

Is it a battery or transformer?

If transformer-fed, what voltage is it supplied with?

Is it an open circuit system or a closed circuit system?

 NEVER examine live equipment ALWAYS switch off first.

Local supply systems

It is often more practical to have special luminaires that have their own power source: self-contained luminaires; these can then be wired into the standard lighting supply circuit (Figure 8.7). These luminaires consist of a small battery-charging unit, batteries, relay and lamp. The batteries are constantly on charge all of the time the mains circuit is working correctly. When the mains supply fails the internal batteries take over.

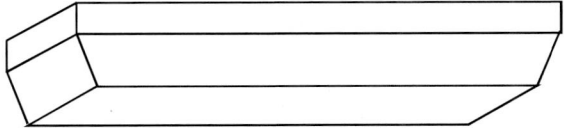

Figure 8.7   *Emergency luminaire with its own batteries and charger*

Maintained lighting

In public areas, such as theatres, the emergency signs have to be illuminated all of the time. Usually in these emergency signs the lamps are supplied by the battery which, under normal conditions, is being constantly charged. When the mains supply fails the batteries continue to keep the lamps illuminated.

Non-maintained lighting

In a non-maintained circuit (Figure 8.8) the lights are only used when the mains supply fails. In luminaires that have their own batteries contained within them a relay switches the lamp on to the battery when the supply fails.

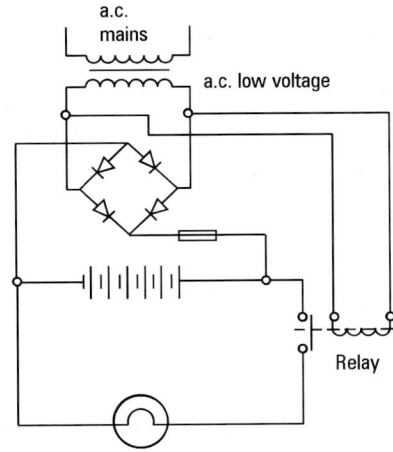

Figure 8.8   *Non-maintained system*

Lighting circuits

The basic lighting circuit is very similar in theory to the basic bell circuit. However, when this has to be installed using two-core cable with circuit protection conductors (cpc), it can become complex. To allow for the fact that twin and cpc cable is not to be split into single core, a three-plate wiring system has been developed. The circuit diagram for this is shown in Figure 8.9 and the practical circuit in Figure 8.10.

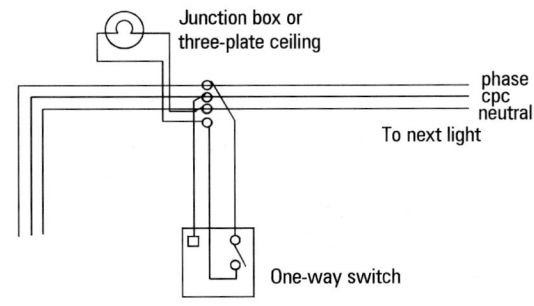

Figure 8.9   *Three-plate lighting circuit diagram*

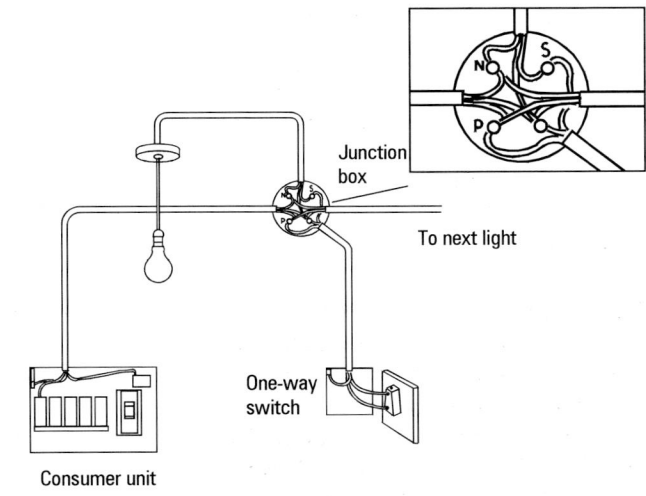

Figure 8.10   *Three-plate junction box*

Two-way switching

On occasions it is necessary to have two switches controlling one light. This often happens on staircases or on landings and halls. In these cases special switches are used together with three-core and cpc cables. The circuit diagram for this is shown in Figure 8.11. This is known as the conversion method, as by the use of three-core cable and changeover switch a one-way light may be readily converted to two-way (Figure 8.12).

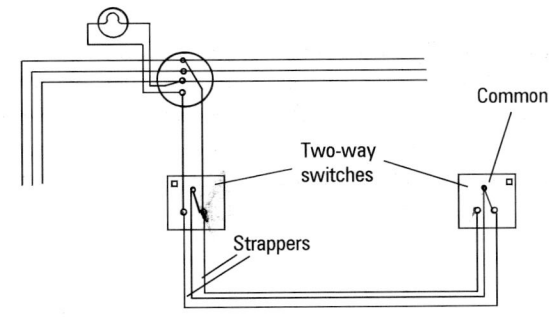

Figure 8.11   *Two-way lighting circuit diagram*

Figure 8.12   *Two-way switch showing connections*

Intermediate switching

Where three switches are required an intermediate circuit is used, as shown in Figure 8.13. A different switch with four terminals has to be used for the intermediate position. The internal switch connections are as shown in Figure 8.14.

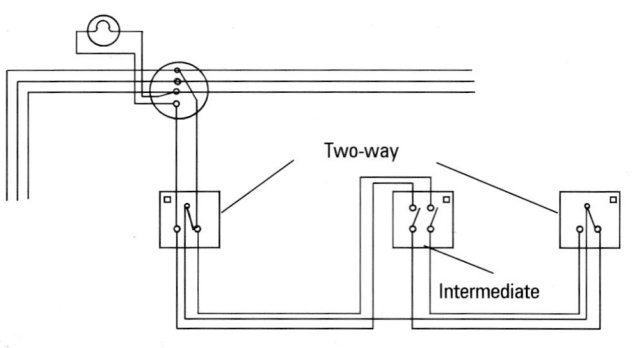

Figure 8.13   *Two-way and intermediate switches circuit diagram*

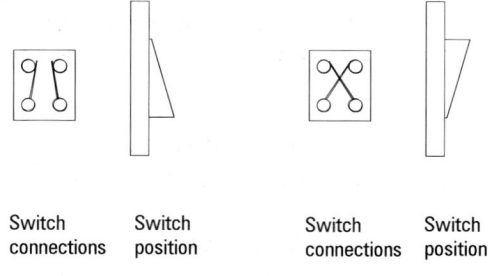

Figure 8.14

Details of how these circuits can be installed can be found in "Practical Requirements and Exercises", another book in this series.

Lighting circuit loading

In normal domestic situations it is sufficient to use a 5 A or 6 A protection device controlling each lighting circuit. This means that the cable used must be capable of carrying this amount of current, and the load required by the circuit should not exceed this. In practical terms this means that on a 5 A circuit supplied with 230 V the maximum power is

$$P = V I \cos \phi$$

$$P = 230 \times 5 \times 1 = 1150 \text{ watts}$$

If we assume a minimum of 100 W is connected to each outlet then there could be up to 11 lamps on that circuit. If the protection device is increased, and the cable rating is suitable, then the number of lamps can be increased (Figure 8.15).

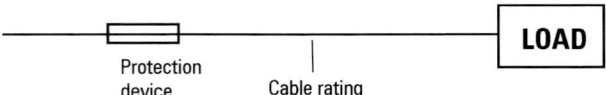

Figure 8.15

Care must be taken to ensure all of these are compatible.

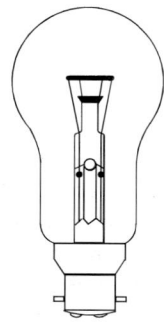

Try this

Without removing or dismantling anything look at a small domestic installation and determine how many wall-controlled one-way switches, two-way switches and intermediate, if any, switches are used.

One-way switches

Two-way switches

Intermediate switches

Types of lamp

The lamps available can be divided into two main groups:
* filament
* discharge

These groups can be further divided to include specific types.

Heated filament

Incandescent

This (Figure 8.16) is the lamp that we are all used to in the home. It consists of a tungsten filament which, when heated up, gives off light.

Figure 8.16   *Incandescent lamp*

Tungsten halogen

This gives off far more light than the incandescent lamp. It consists of a tungsten filament in a thin glass tube which contains a halogen gas (Figure 8.17). It is very important that these are installed to the manufacturer's instructions as they get very hot.

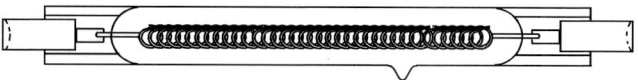

Figure 8.17   Tungsten halogen lamp

Extra low voltage (**ELV**) tungsten halogen lamps are used extensively for domestic lighting. They are also used with an integral transformer for display spotlights.

Discharge

In a discharge lamp the light is produced by current flowing through a gas or vapour. To create and control the discharge that has to take place to produce light, special control gear has to be used. This is usually based on a choke or transformer which has a high inductance. Equipment with inductance can draw high currents, due to the effect of power factor, unless it is corrected. In most cases a capacitor is connected across the supply to the luminaire. This keeps the current demand to a minimum to limit the rating of circuit conductors.

The fluorescent fitting is the most widely used discharge lamp, and Figure 8.18 shows a typical circuit diagram.

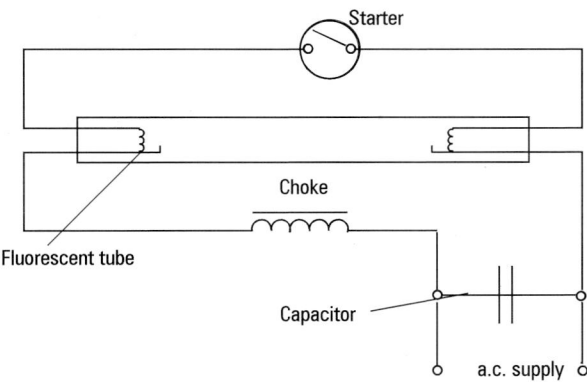

Figure 8.18

The circuit consists of a choke, a starter, a low-pressure mercury vapour lamp (fluorescent) and a power factor correction capacitor. When the circuit is first switched on current flows through the choke, one filament of the discharge tube, through the starter switch which is closed, through the other filament and back to the supply. As this is an alternating current the magnetic field is continuously building and collapsing in the iron core of the choke.

The contacts in the starter switch open due to a temperature rise in the starter, and as the contacts open the sudden collapse of the field in the choke applies a high voltage across the discharge tube. The starter switch is now shorted out by the tube, so it remains out of the circuit. The choke now becomes a current limiting device and keeps the lamp current down to an acceptable level.

If the choke were to short-circuit the lamp current would rise to an unmanageable level and the tube filament could blow. The power factor correction capacitor across the supply keeps the circuit currents in phase with the voltage so that the current in the circuit is kept to a minimum.

Other low voltage (LV) fluorescent fittings are also available.

Switches

When a fluorescent, or any other discharge lamp, is switched off, a high voltage is induced across the switch contacts. For this reason suitable switches have to be used for all discharge lighting circuits.

Stroboscopic effect

One of the problems with using discharge lighting is what is known as the "stroboscopic effect". The current across a discharge tube is continually changing. If this is the only light source over rotating machinery the speed of the rotation of the machine can *appear* to be slowed down or even stationary. This, of course, can be very dangerous, and precautions must be taken to avoid it.

The stroboscopic effect can be reduced by

- connecting adjacent fluorescent luminaires on different phases
- using special lead/lag luminaires
- having a high-power tungsten lamp shining on moving parts
- using fluorescent luminaires operating on high-frequency circuits.

Try this
In a domestic installation, without disconnecting or removing anything, determine and list the types and power rating of lamps in use.

Type	Power rating

Points to remember ◄ – – – – – – – – – – – – –

When a bell circuit is used as an alarm circuit it is usually a "closed circuit system".

Emergency lighting systems usually rely on batteries when the main supply fails. These are generally secondary cells either located centrally or within each luminaire. Standby generators may be used in very large installations.

The cable used in lighting circuits must be capable of carrying the load required. The lamps available for lighting can be divided into two main groups, heated filament and discharge.

Try this

1. What is the maximum power on a 13 A circuit supplied with 230 V?

2. What is the maximum power that can be connected to a 230 V circuit protected by a 30 A protection device?

Self-assessment multi-choice questions
Circle the correct answers in the grid below.

1. (i) On an "open circuit" alarm system an extra push must be connected in series with the first.
   (ii) On a "closed circuit" alarm system an extra push must be connected in series with the first.
   (a) statement (i) is correct and statement (ii) incorrect
   (b) statement (i) is incorrect and statement (ii) correct
   (c) both statements are correct
   (d) both statements are incorrect

2. When a light is to be switched by three switches one of the switches will be
   (a) triple pole
   (b) two-way
   (c) intermediate
   (d) double pole

3. The capacitor which is fitted to a fluorescent light circuit has the effect of
   (a) causing the lamp to light more quickly
   (b) drawing more current from the supply if connected
   (c) drawing less current from the supply if connected
   (d) extending lamp life

4. The stroboscopic effect on rotating machines can occur where which type of lighting is installed?
   (a) incandescent
   (b) tungsten halogen
   (c) extra low voltage
   (d) discharge

5. Stroboscopic effects can be reduced by
   (a) connecting adjacent luminaires on different phases
   (b) increasing the supply voltage
   (c) reducing the supply voltage
   (d) fitting resistive ballasts

Answer grid

1	a	b	c	d
2	a	b	c	d
3	a	b	c	d
4	a	b	c	d
5	a	b	c	d

Part 2

Power circuits

The ring final circuit

The most common socket outlet circuit used in domestic installations is the ring final circuit (Figure 8.19). As the name implies, the circuit starts and finishes at the same point and should form a continuous ring. The phase starts and finishes at the same protective device (fuse or circuit breaker), the neutral starts and finishes at the same connection on the neutral bar, and the circuit protective conductor starts and finishes at the same connection on the earth block. Each socket outlet on the ring has at least two cables to it. This means that there must be at least two conductors in each terminal.

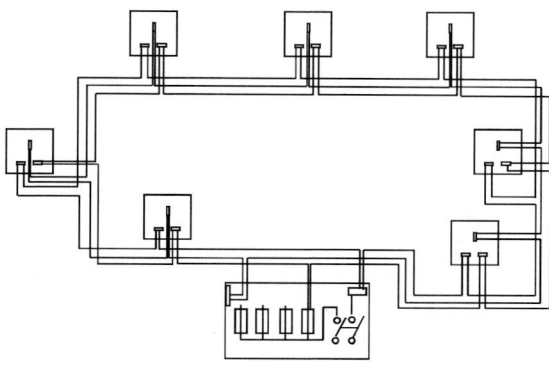

Figure 8.19   A ring final circuit

In general domestic premises there should be a separate ring final circuit for every 100 m^2 of floor area. We must remember that the maximum load that can be connected to a ring protected by a 30 A or 32 A fuse is just over 7 kW, and the number of circuits selected accordingly.

Remember
The bare circuit protective conductors should be sleeved with green/yellow sleeving

Non-fused spurs

Where a socket is required at a point away from the run of the ring circuit cables a non-fused spur (Figures 8.20–8.22) may be used. This is a single cable run just to this outlet. The cable conductors for a non-fused spur must not be of a smaller cross-sectional area than that of the ring conductors. Usually in domestic premises a 2.5 mm^2 conductor is used with a 1.5 mm^2 circuit protective conductor.

The non-fused spur may be connected from
- a socket on the ring
- a junction box connected onto the ring or
- the consumer unit or distribution board

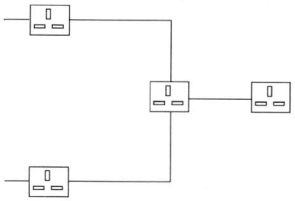

Figure 8.20   Non-fused spur from a 13A socket outlet

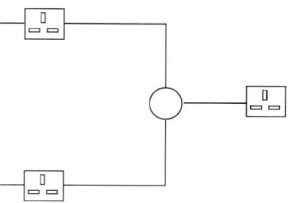

Figure 8.21   Non-fused spur from a junction box

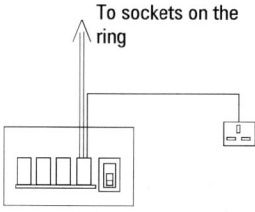

Figure 8.22   Non-fused spur from a distribution board

A spur must not be more than one single or one twin socket outlet.

Radial circuits

Radial circuits, unlike ring circuits, do not return back to the supply point (Figure 8.23). They are basically a loop in and out circuit. As the current distribution is not as good as the ring circuit there are tighter limitations on its use.

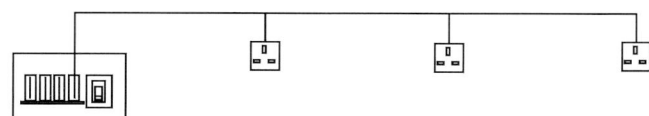

Figure 8.23

Cooker circuits

Cooker circuits are usually a single outlet supplied from a separate protection device in the distribution board. The outlet is normally a double pole switch which controls the supply to the cooker. The rating of the protection device and cable is determined on the basis that not all of the cooker will be switched on at the same time. Even once everything is switched on the thermal control devices, such as thermostats and simmerstats, would be continually switching sections off. Making allowance for this process is known as applying diversity. So that calculations can be made as to the possible load of a domestic cooker a formula has been devised.

This works on the **first 10 amperes** of the total possible being taken at **100%**

the **remainder**, after the 10 amperes are taken off, are taken at **30%**

plus 5 amperes if a **socket outlet** is incorporated in the control unit.

Example

The assumed demand for a cooker which contains:

2×1.5 kW hob plates

2×3.0 kW hob plates

1×2.0 kW oven/grill

1×4.0 kW oven

has a maximum total power of 15.0 kW.

The maximum current would be

$$I \quad = \frac{P}{V} = \frac{15 \times 1000}{230}$$

$$= 65.22 \text{ A}$$

The assumed current demand, allowing for diversity, is

the first 10 amperes are at 100%	=	10 A
that leaves 55.22 amperes at 30%	=	16.57 A
total	=	26.57 A

This means that the cable supplying this cooker would have to have a rating of at least 26.57 A. If the control unit contained a socket outlet the rating would have to be at least 31.57 A.

Water heating

Unlike cookers, there is no reduction off the total rating on water heaters. They are either ON or OFF, full load or no load, even though they are thermostatically controlled. It is usual to connect immersion heaters (Figure 8.24) to their own circuit, as they are often rated at 3 kW.

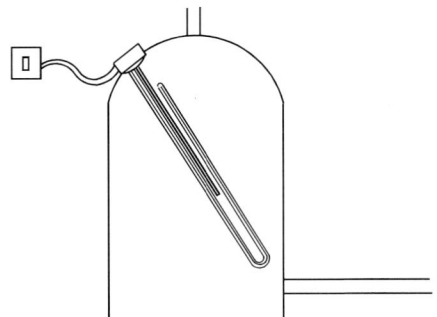

Figure 8.24   Immersion heater

Instantaneous water heaters

Instantaneous water heaters with immersed and uninsulated elements must be permanently connected to the electricity supply through a double-pole linked switch. Plugs and socket outlets must *not* be used.

Space heating

Total electric space heating generally makes use of low-cost electricity at night. These systems are controlled by tele-switches on tariffs such as Economy 7, which charges customers about half price for electricity used over a particular seven hours at night. Such heating devices have to be designed so that heat produced at night can be stored and given off during the day.

Special "storage heaters" have been developed that consist of elements embedded in refracting bricks (Figure 8.25). The elements heat the bricks up at night and they retain the heat and give it out as required. To help control the heat being given out, thermal lagging is placed all around the brick core. The amount of heat being put into the heater is controlled by a thermostat fitted inside the case.

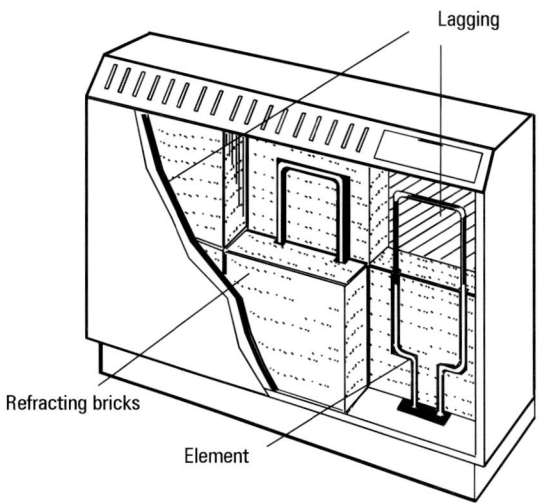

Figure 8.25   *Storage radiator*

Motor circuits

For most industrial purposes motor circuits are usually three-phase, whilst for domestic and office use single phase supplies are more common.

BS 7671 requires the provision of an isolating device to allow the disconnection of the supply and a method of preventing the automatic restarting of a motor after a stoppage or failure of supply. Additionally, every motor over 0.37 kW must be provided with a means of overload protection.

All these requirements can be included in a motor starter (Figure 8.26), although a separate starter and isolator are often used.

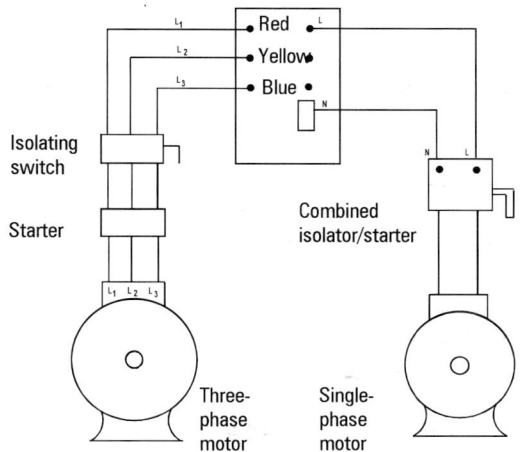

Figure 8.26

Some three-phase motors have windings connected in star/delta for easier starting purposes and six connections are necessary between the motor and starter (Figure 8.27).

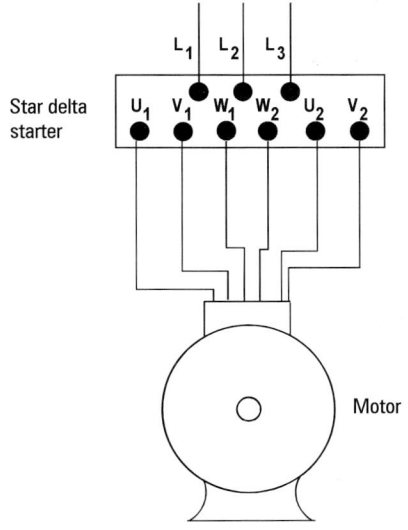

Figure 8.27

Temperature control devices

Cookers and water heaters use special control devices to regulate temperature. Thermostats and simmerstats have been covered in the book *Basic Science and Electronics* but we will just remind you of the basic principles before we look at the three heat switch.

Thermostats

These usually work on the expansion of gas, liquid or metal depending on their use.

Expansion of gas or liquid

An oven thermostat has a bulb of gas or liquid that is placed in the oven and connected to the control switch by a capillary tube. This is very similar to that used in a refrigerator or freezer but the switch action is reversed.

Expansion of metals

Thermostats are also used for room heating to control the boiler pump or space heater. The active section here is a bimetal strip which, when there is a temperature change, bends or straightens out. The thermostat in an immersion heater also works on the expansion of metals. A rod thermostat slides down inside a tube into the water for accurate measurements. These work on the principle of the brass outer tube expanding at a greater rate than the inner rod. The result of the different lengths operates the switch contacts.

Simmerstats

The simmerstat has its own source of heat which controls the action of the bimetal strip. These can be very useful to limit the heat of a hot plate and the control can have a number marking to give an indication of how high or low the setting is.

Remember

The type of temperature control device used must be appropriate for the control required and the location.

The thermostat controls the output of a heating device by measuring the temperature and cutting off the supply when the required temperature is reached.

The simmerstat controls the energy input to the device and has no reference to external temperature.

The three-heat switch

This is a method of connecting two identical elements so that three variations of output can be obtained. The switch is made so that the internal connections take up new positions when the control is rotated. The circuit when connected to the switch is as in Figure 8.28.

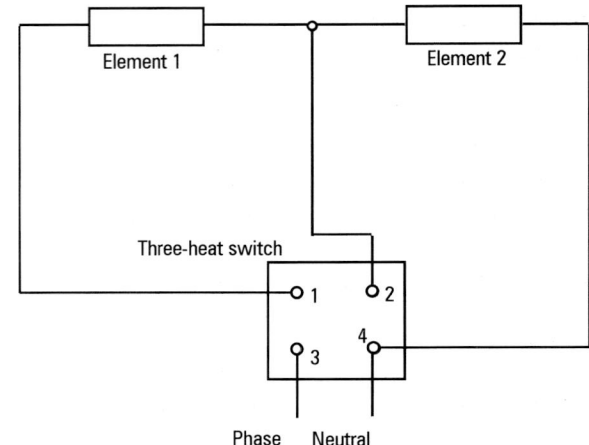

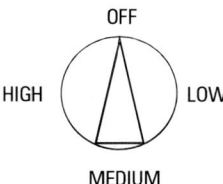

Figure 8.28

The elements are connected as follows

Switch position	Element connections	Switch connections
OFF	No circuit	
LOW	The elements in series	1 to 3
MEDIUM	One element connected	2 to 3
HIGH	Both elements in parallel	2 to 3 and 1 to 4

The heats given off go in direct proportions – LOW is half of MEDIUM and MEDIUM is half of HIGH.

Special conditions

We have seen how circuits can be connected and controlled. We must remember that electricity can be very dangerous and must always be kept under control. A particularly hazardous room in the domestic installation is the bathroom (Figure 8.29). There is always water and condensation about, and in many cases exposed metalwork. In simple terms, when skin becomes damp it becomes a better conductor of electricity. In a bathroom, due to the very nature of the room's use, people are at a greater risk of electric shock, which could prove fatal.

For this reason the bathroom has special regulations that apply. These in general terms mean that

- no general socket outlets should be installed
- no switches should be installed within the reach of persons using the bath or shower
- all lamps within 2.5 m of the bath or shower should be constructed of or shrouded in insulating material; alternatively totally enclosed luminaires should be used
- heaters with exposed elements should not be installed within the reach of persons using the bath or shower

For every one of these there are *exceptions*:

- socket outlets complying to BS 3535 for shavers can be installed
- there can, under some special circumstances, be 12 V socket outlets installed
- switches of the pull-cord type can be installed providing the switch body is out of reach of persons using the bath or shower
- lamps can be unenclosed if they are in the appropriate skirted lampholders complying to BS 5042 (or BS EN 61184) or at least 2.5 metres from the bath or shower
- heaters with exposed elements can be installed providing they cannot be touched from the bath. Incidentally, a silica tube element is classed as being exposed

In addition to these requirements, all exposed metalwork, that is metalwork that can be touched, which is not normally live but which may become live under fault conditions, must be bonded together within the bathroom to form an equipotential zone.

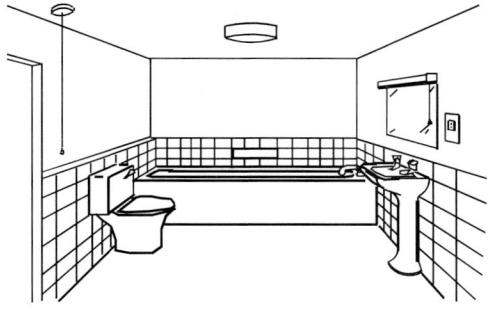

Figure 8.29   The bathroom

It is not only bathrooms that contain showers, so special precautions have been introduced to allow for this (Figure 8.30). For example, socket outlets should be at least 2.5 metres away from the shower cubicle when it is in a room which is not in a bathroom.

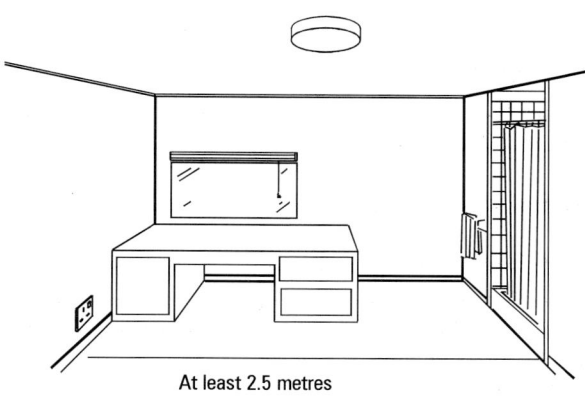

At least 2.5 metres

Figure 8.30   A shower cubicle in a room other than a bathroom.

Remember

In bathroom installations special regulations apply.
Reference should be made to BS 7671 for the particular additional requirements for rooms containing a bath or shower.

Points to remember ◄ — — — — — — — — — — — —

Most domestic socket outlets are supplied from ring final circuits, but sometimes it is more convenient to use radial circuits.

We need to determine the load to be connected, then select a protective device based on that load.

Cooker circuits can be sized using diversity to determine the load.

No diversity is allowed for water heater or space heater circuits.

Bathrooms require special safety considerations.

Self-assessment multi-choice questions
Circle the correct answers in the grid below.

1. In general terms a separate ring circuit protected by a 30 A or 32 A protection device should be installed for every
    (a) 20 m^2 of floor space
    (b) 50 m^2 of floor space
    (c) 100 m^2 of floor space
    (d) 200 m^2 of floor space

2. What is the assumed current demand of a socket outlet to BS 1363?
    (a) 13 A
    (b) 16 A
    (c) 20 A
    (d) 32 A

3. What is the maximum total connected load of a cooker which contains:
    1 × 1.5 kW hob plate
    2 × 2 kW hob plates
    1 × 4 kW oven/grill
    (a) 7.5 kW
    (b) 9.5 kW
    (c) 11.5 kW
    (d) 13.5 kW

4. If the cooker in question 3 is connected to a 230 V supply, and there is no socket outlet involved, the assumed current demand, allowing for diversity, is
    (a) 9.5 A
    (b) 19.39 A
    (c) 29.39 A
    (d) 39.39 A

5. How much greater is the heat given off when a three heat switch is in the high position compared to the low setting?
    (a) half as much again
    (b) twice as great
    (c) 4 times as great
    (d) 8 times as great

6. How far from a bath or shower can an unenclosed lamp be installed without an appropriate skirted lampholder?
    (a) 1.5 m
    (b) 2.0 m
    (c) 2.5 m
    (d) 3.0 m

7. The socket shown in Figure 8.31 is connected as a

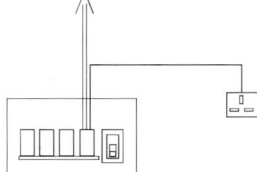

Figure 8.31

    (a) radial spur
    (b) fused spur
    (c) non-fused spur
    (d) socket on a ring circuit

8. Protection against overload must be incorporated in every motor control when the rating of the motor exceeds
    (a) 0.3 kW
    (b) 0.37 kW
    (c) 0.7 kW
    (d) 1.3 kW

9. *All* circuits to general purpose electric motors must include
    (a) a plug or socket outlet
    (b) an emergency stop
    (c) a means to prevent automatic restarting
    (d) overload protection

10. An appropriate type of temperature control used to control a domestic refrigerator would be
    (a) invar rod
    (b) gas filled
    (c) bi-metal strip
    (d) three heat switch

Answer grid

1	a	b	c	d		6	a	b	c	d
2	a	b	c	d		7	a	b	c	d
3	a	b	c	d		8	a	b	c	d
4	a	b	c	d		9	a	b	c	d
5	a	b	c	d		10	a	b	c	d

9

Protection

Complete the following to remind yourself of important facts from the last chapter.

There are two ways in which electricity is used to produce light. The one which uses a heated _____ is called _____

In the other method, an electric _____ passes through a gas or vapour.

An economic means of heating a building by electricity can be obtained by means of the _____ tariff. The supply from the Regional Electricity Company is controlled by _____

The heating elements of a storage radiator are surrounded by _____

_____ of appropriate types are used to control heating loads.

Domestic socket outlets are generally installed as a _____ circuit, although _____ circuits may sometimes be more appropriate.

No allowance for diversity can be made to a circuit supplying a _____ heater. Cooker circuit diversity is applied as the first _____, plus _____% of the remainder plus ____ if a socket outlet is fitted in the cooker control panel.

Because of the _____ _____ risk _____ are subject to additional special requirements.

On completion of this chapter you should be able to:

◆ state the general requirements for safety in regard to isolation and protection of an installation
◆ give examples of where protection is incorporated at a main intake position
◆ explain how socket outlets could become overloaded
◆ describe a short circuit
◆ show by a simple sketch what is meant by discrimination
◆ explain why it is necessary to connect all installations to earth
◆ complete the revision exercise at the beginning of the next chapter

Part 1

The main requirements

There are at least four different factors that must be considered when dealing with control and protection of an installation.

These are:
• isolation and switching
• protection against thermal effects
• protection against overcurrent
• protection against indirect contact

These factors are all concerned with safety and need to be installed for protection, not just from electric shock, but also from fire, burns or injury from mechanical movement of equipment which is electrically activated.

Isolation and switching

The term isolation, within this context, means the cutting off of the installation, or circuit, from all sources of electrical supply to prevent danger.

In a domestic installation the main means of isolation is often the switch controlling the consumer unit (Figure 9.1).

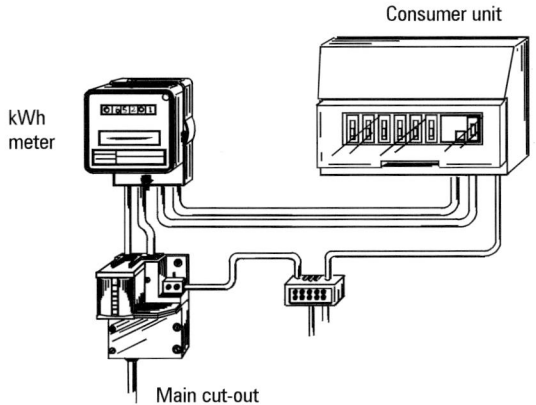

Figure 9.1   *The main intake position of a domestic installation*

Within the installation other local methods are used and the positioning and type of these can be very important.

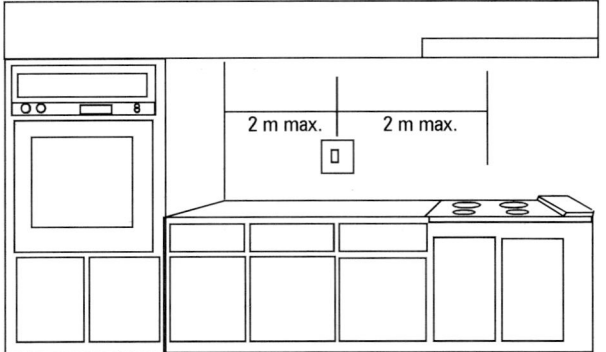

Figure 9.2

The control for cookers is a D.P. switch which should be positioned no further than 2 m from the cooker or, as in Figure 9.2, from either part of the cooker (IEE Guidance Note 1). Figures 9.3 and 9.4 show the arrangements for immersion heaters and domestic boilers.

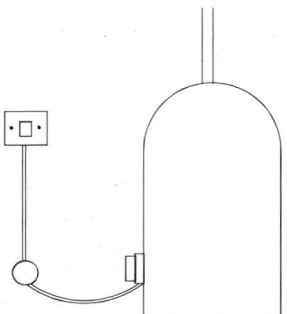

Figure 9.3   Immersion heaters must have a DP control switch adjacent to the heater

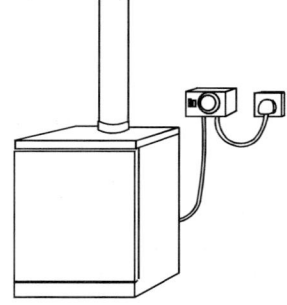

Figure 9.4   Domestic boilers often have a plug and socket as the means of isolation.

In an industrial installation this can become far more complex, as you can see in Figure 9.5. Whereas in a domestic installation most isolators are double-pole (D.P.) types, in industrial situations there will also be triple-pole (T.P.) and triple-pole and neutral (T.P.& N.). Some examples of the use of isolators in industrial installations are shown in Figures 9.6 and 9.7. Figure 9.8 is an example of emergency switching.

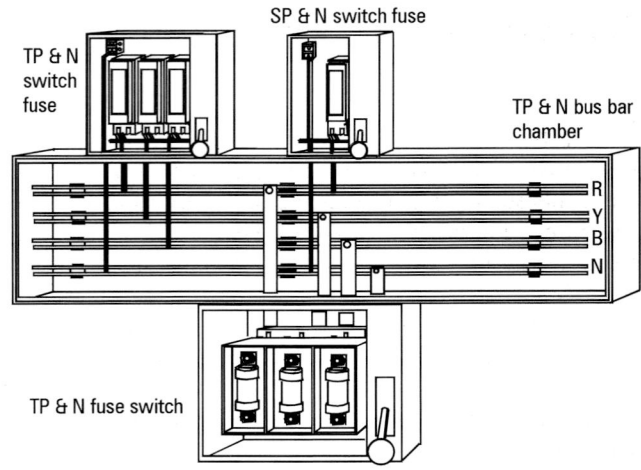

Figure 9.5   Industrial distribution

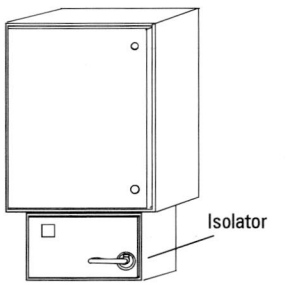

Figure 9.6   Three-phase distribution board controlled by an isolator

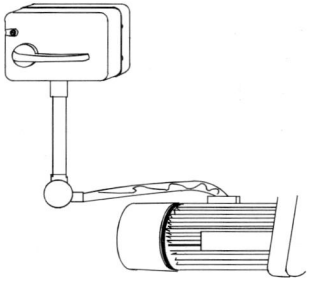

Figure 9.7   Motor with isolator adjacent

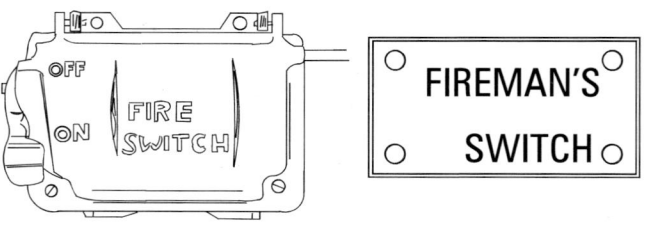

Figure 9.8   Fireman's emergency switch

Remember
An isolator must cut off an electrical installation, or part of it, from every source of electrical energy.

Overcurrent protection

Overload

An overload is a situation that occurs in a circuit which is still electrically sound. It is generally caused by trying to take more power from a circuit than it is designed for. This results in a larger than normal current flowing in the circuit. If the load is reduced then the circuit can continue to function without any need for repair. A typical example of this is a radial circuit which has been extended and the load has increased to exceed that originally intended (Figure 9.9). Each item connected to the circuit and the circuit supplying the equipment is healthy, and so is the circuit supplying the outlet, but more current is being drawn through the cable than was originally intended.

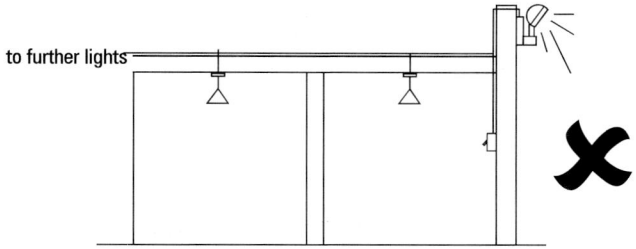

Figure 9.9   *A radial circuit with 10 existing lights has a 500 W tungsten halogen lamp added which increases the load to exceed that originally intended.*

Every cable has some resistance, and the result of drawing more current through the cable results in more heat being produced in the cable. This rise in temperature will, over a period of time, result in the insulation becoming less effective and eventually breaking down. In the case of severe overload the insulation becomes so hot it begins to melt and may even catch fire. This is obviously a serious fire risk (Figure 9.10) and we must take steps to prevent this happening.

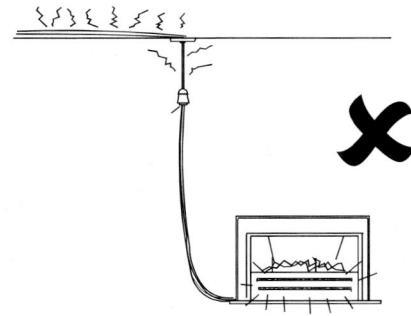

Figure 9.10

Protection devices

The usual way of protecting each cable is by installing a device in each circuit that will automatically disconnect it from the supply when an overload occurs. There are several different protection devices that can be used, one of which is the rewireable semi-enclosed fuse (BS 3036) (Figure 9.11). However, these devices are not the most effective, and their use requires special consideration to be made for cable sizing.

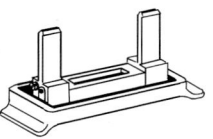

Figure 9.11   *Semi-enclosed fuse BS 3036*

Other devices are the cartridge fuses (BS 1361), including high breaking capacity fuses with BS 1362 for plug tops, fused spurs and the like, BS 88 part 2 & part 6, and circuit breakers (BS EN 60898) (Figure 9.12).

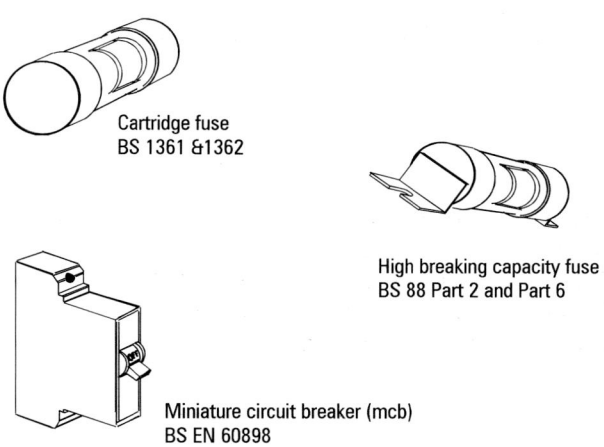

Figure 9.12

Overloads are sometimes difficult for a device to detect, as they can build up over a period of time as different pieces of equipment are switched on. For example, a lighting circuit is usually designed for about ten 100 W lamps. If each 100 W lamp was replaced by a 150 W lamp and they were all switched on one after the other until they were all on, the circuit would eventually be carrying one and a half times the load it was designed for. It would depend on the protection device as to whether the circuit would be automatically disconnected under these conditions.

Table 9.1

Examples of current ratings for protection devices

Protection device	Current ratings								
Fuses to BS 3036	5	15	20	30	45				
Fuses to BS 1361	5	15	20	30	45				
Fuses to BS 88 Pt2 & Pt6	6	10	16	20	25	32	40	50	63
Circuit breakers to BS EN 60898	6	10	16	20	25	32	40	45	50	63

Short circuit

A short circuit is said to be a connection of negligible impedance between live conductors. "Live conductors" means all those carrying current under normal conditions, which includes the neutral conductor.

If a short circuit is two conductors touching then it can be assumed that the resistance or impedance of that connection would be so low that it could be ignored.

So a short circuit can occur between:

- conductors connected to different phases for example, red phase to yellow phase or red phase to blue phase,
- any phase and neutral

Let us consider for a moment the implications of this. If the fault has negligible impedance then the only restriction to the amount of current that will flow in the circuit is that of all the conductors. As conductors are of a low resistance then this total value will itself be low and the current that flows can be very high. If we look at Figure 9.13 we can see that the impedance of the total circuit under these conditions is made up of

the supply transformer windings (0.01 Ω)

the supply cable, both phases and neutral
(0.01 Ω + 0.01 Ω)

the cables up to the short circuit
(0.15 Ω + 0.15 Ω)

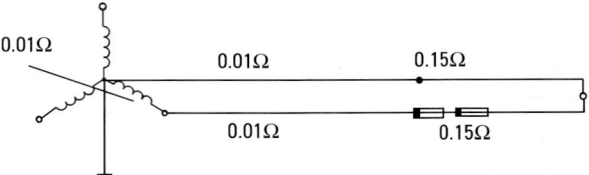

Figure 9.13

This gives us a total impedance of:
0.01 + 0.01 + 0.01 + 0.15 + 0.15 = 0.33 Ω

If the system was operating at 230 volts then the current flow under these fault conditions would be 230 volts divided by 0.33 ohms

$$\frac{230\ V}{0.33\ \Omega} = 697\ \text{amperes}$$

Considering that this circuit may be protected by a 30 A fuse or circuit breaker this is a large current to flow in a domestic installation.

Not only is the current flow going to be high, it also tends to occur very quickly, so the protective device has to be able to cope with a very different set of conditions from those of an overload. In this case the temperature builds up very rapidly, within fractions of a second, and the device must sense this current and disconnect it from the supply before any damage is done to the cable or equipment. The risk of fire under these conditions is considerable.

Now, if a fault should develop at the intake position of an installation the limiting impedance is only that of the supply conductors. Taking the example in Figure 9.14, the external impedance to the installation would be

0.01 + 0.01 + 0.01 = 0.03 ohms

this means that on a 230 volt supply the fault current would be

$$\frac{230V}{0.03\Omega} = 7666.6\ \text{amperes or 7.6 kA}$$

This value is known as the prospective short circuit current.

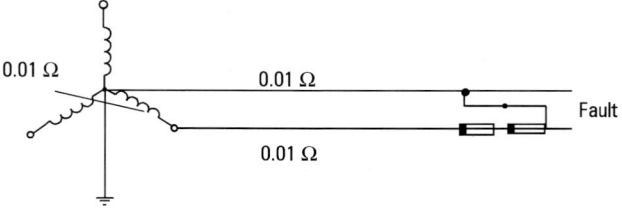

Figure 9.14

Protection equipment fitted at any point needs to be capable of breaking the prospective short circuit current, which may occur at that point, without damage to the equipment.

Discrimination

In any installation it is not possible to protect all circuits and equipment using a single device. Protection equipment designed to protect a circuit with a load of 5 amperes will not serve as protection for a circuit rated at 30 amperes, and vice versa. A typical arrangement of an internal system to cope with this situation is shown in Figure 9.15, and we can see that there are a number of fuses between the supply intake cable and the final load of the table lamp on the socket outlet circuit.

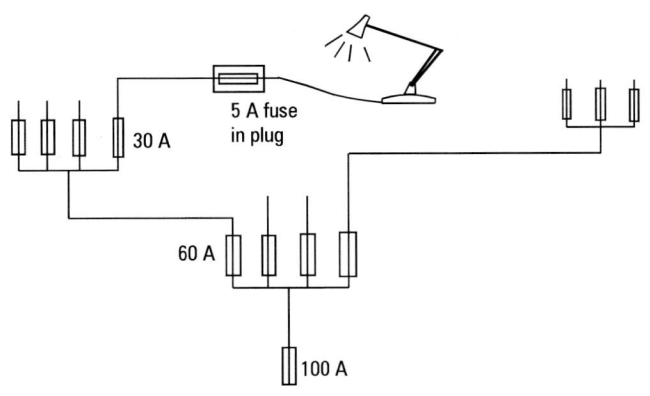

Figure 9.15

The objective of discrimination is to make sure that an overcurrent occurring at any point on the system causes the minimum disruption of supply. To do this we must ensure that the fuse on the supply side closest to the cause of overcurrent operates first and leaves the other devices intact, thus minimising the number of circuits or appliances affected. So should a fault occur on our table lamp only the lamp should be disconnected from the supply by the plug fuse, leaving everything else functioning normally.

Let us look at a couple of examples of discrimination in practice.

In Figure 9.16 the design for this part of a system has a 45 A fuse supplying a distribution board, which supplies three circuits, one of 45 A, one of 15 A and one of 5 A. Should a fault occur on the 45 A circuit, the fuse supplying the distribution board may operate and all three final circuits would then be disconnected from the supply.

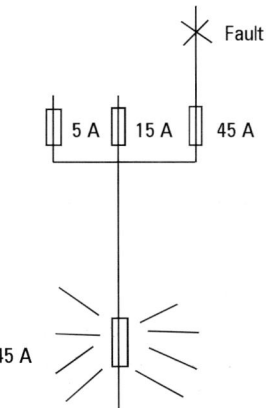

Figure 9.16

In the Figure 9.17 design the 45 A fuse supplying the board has been uprated to 60 A. Now should the same fault occur on the 45 A circuit the circuit fuse will operate, leaving the other two circuits unaffected. The cables may now be selected based upon the sizes of protective devices required to provide discrimination.

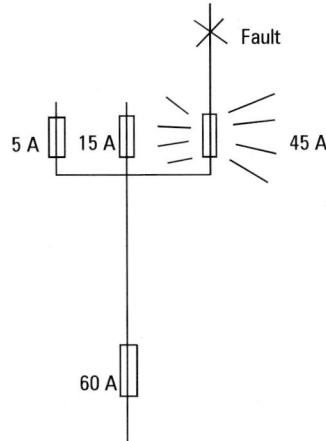

Figure 9.17

In practice, the loading of all circuits should be considered before the installation begins so that the correct rating and type of protective devices can be installed. To change the protective device for one of another type or rating after completion may require extensive alterations.

Try this

A 1 kW electric fire is plugged into a 13 A socket. The socket is protected by a 30 A fuse in a consumer unit. The main fuse protecting the supply cable and backing up the consumer unit is 100 A. Draw a line diagram similar to those in Figures 9.15, 9.16 and 9.17 showing the fuses and state which fuse will operate if the 13 A socket develops a short circuit.

Protection against thermal effects

People, fixed equipment and materials that can come into contact with electrical equipment must be protected against any harmful effects of heat such as fire or burns.

Points to remember ◄ – – – – – – – – – – – – – –

At the _____ position of a standard domestic installation there should be some method of _____ the consumer's circuits completely from the incoming _____. There should also be a device that will _____ _____ in the event of an overload. Overcurrents can be caused by either _____ or _____.

An overload occurs in a circuit which is electrically sound.

An overload may develop over a period of time and the rate at which the current rises may be slow.

Short circuits result in a very rapid increase in current to a particularly high level.

Protective devices are manufactured to a number of _____ _____, each having a different

_____.

When selecting a _____ _____ for a particular circuit we must ensure that the _____ type is used. The characteristics for both overload and short circuit should be considered.

When proposing an addition to an electrical system it is important to ensure that the _____ system has sufficient capacity to allow additional loads to be connected. This applies to all parts of the system: main supply, distribution boards, cables and fixed circuits.

When a fault occurs it is important that the protection device closest to the fault on the supply side operates. This helps to ensure that equipment is not disconnected unnecessarily by disconnecting only the faulty circuit and leaving the others unaffected.

Self-assessment multi-choice questions

Circle the correct answers in the grid below.

1. A requirement for safety that all installations should have is:
   (a) fire alarm
   (b) overcurrent protection
   (c) indication of supply failure
   (d) energy meter

2. The term "isolation" means
   (a) separated from all other circuits
   (b) in a separate building
   (c) mounted in a separate enclosure
   (d) cut off from all sources of electrical supply

3. An example of an overloaded circuit is a
   (a) 13 A socket with a 2 kW electric fire connected
   (b) lighting circuit with a 100 W lamp connected
   (c) 13 A socket outlet with 100 W lamp connected
   (d) lighting circuit with a 2 kW fire connected

4. The consumer's isolator in a new domestic installation is most likely to be located
   (a) on the socket outlets
   (b) in the consumer unit
   (c) in the kWh meter
   (d) in the cut-out

5. TP & N stands for
   (a) three phase and negative
   (b) two pole and neutral
   (c) triple pole and neutral
   (d) three pin and negative

6. An overload occurs
   (a) in a circuit that is electrically sound
   (b) between two conductors at the same potential
   (c) between two phase conductors at different potentials
   (d) between a live conductor and earth

7. A short circuit occurs
   (a) in a circuit that is electrically sound
   (b) between two conductors at the same potential
   (c) between two phase conductors at different potentials
   (d) between two conductors on the same phase

8. Discrimination when applied to fuses means to
   (a) fit the correct make of fuse
   (b) make certain that all the fuses are of different sizes
   (c) ensure the correct sequence of operation of fuses
   (d) ensure the fuse nearest to the source of supply operates first

9. If a fault occurs at the point shown in Figure 9.18, discrimination is said to have taken place if which protective device operates?

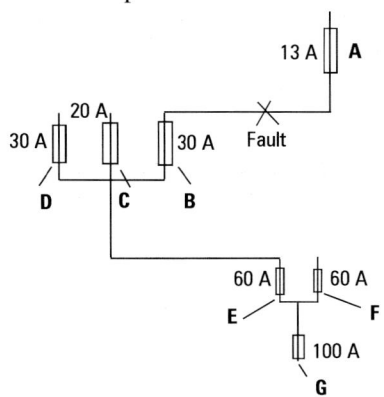

Figure 9.18

   (a) A
   (b) B
   (c) E
   (d) G

10. The short-circuit current flowing through the circuit in Figure 9.19, if the open circuit voltage of the transformer is 230 V, will be

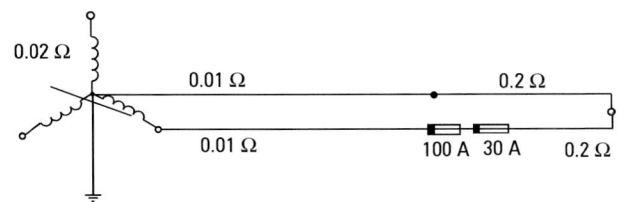

Figure 9.19

   (a) 818.2 A
   (b) 652.7 A
   (c) 522.7 A
   (d) 295.4 A

Answer grid

1	a	b	c	d		6	a	b	c	d
2	a	b	c	d		7	a	b	c	d
3	a	b	c	d		8	a	b	c	d
4	a	b	c	d		9	a	b	c	d
5	a	b	c	d		10	a	b	c	d

Part 2

Earthing

This is a subject that has been mentioned briefly many times. Here we will look at why and how we earth.

Why do we earth?

It is generally accepted that the earth we walk and build on is a conductor. We have no choice over this, so we need to consider it very seriously.

The mass of earth is generally accepted as being of zero volts, and, as our electrical installations operate at voltages above zero, there is a potential difference between the earth and our electrical system. In order that we may control any current flow which may occur to the mass of earth we need to connect it to our system and provide a reference for the earth.

So that the general mass of earth is given a reference, all electricity substations have the star (neutral) point connected to earth. It is also a requirement that all installations have a connection to earth, apart from the neutral conductor. The consumer's earth connection may be provided by the supply company. This in turn may be connected to the metal sheath of the supply cable (TN-S) or to the neutral of the supplier's network (TN-C-S).

If the consumer's earth connection is not provided by the supplier then an independent earth electrode must be installed (TT). In this case it is normal practice to fit a residual current device (RCD).

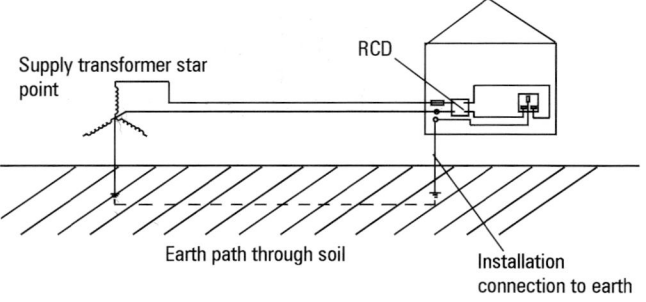

Figure 9.20   TT system

Let us now consider Figure 9.20, where the supply transformer is represented by the star-connected winding.

In this case the consumer's installation is connected to the supply transformer with a phase and neutral connection and the earth connection at the consumer's premises is by the supply cable sheath. An appliance plugged into the socket outlet would be connected between phase and neutral to make it work normally, and if it had a metal case this would be connected to the earth pin of the plug. Under normal working conditions no current would flow in the earthed conductor. However, if a fault developed so that some current went between the phase and the metal case, this fault current must be cleared before it can create a danger.

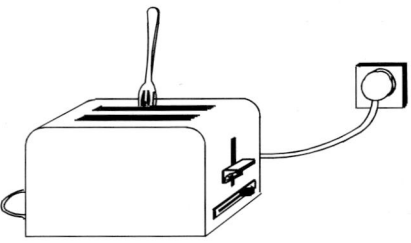

Figure 9.21

As the metal case (in Figure 9.21) is connected to earth the fault current will go down into the earth and flow back to the earthed point of the supply transformer (Figure 9.22).

Remember that it cannot go anywhere else, for there has to be a complete circuit for current to flow through and this is the only complete circuit.

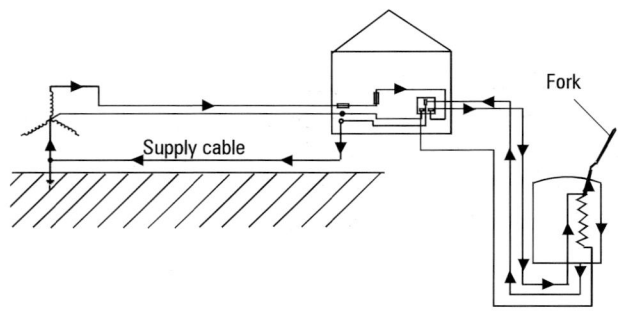

The fork completes a path to earth from the phase conductor.

Figure 9.22   TN-S system

It is important that the earth fault path, as it is known, is kept at as low a resistance as possible, so that in the event of a fault the maximum current can flow sufficient to operate the protective device and disconnect the supply.

However, if this fault has a high impedance the current may continue to flow without the protective device operating.

If, for example, only 40 mA flows to earth on a circuit which normally takes 10 A the current in the phase conductor is only 10 A + 40 mA, and this will not affect the protection device.

It is important to remember that, whilst a current flow of 40 mA appears small enough to ignore, the human body can be affected by currents as low as 10 mA or less. If the fault was through the body of a person there could be a risk of electric shock even at these very low current levels.

Earth fault impedance test

To ensure that this earth fault path has a low enough value a test should be carried out at regular intervals. The test is known as the earth fault loop impedance test; impedance because it is carried out using the a.c. supply. The test injects a current through the earth return path from the premises to the supply transformer, through the windings and back to the consumer's premises through the phase conductor.

The test is easily carried out from a socket outlet by plugging the test instrument in. Care should be taken to ensure that the warning indicators on the test equipment are observed, as the test current could be fatal if a person was to become part of the circuit.

This is one of a number of such tests which need to be carried out as we need to check earth fault loop impedance at the origin of the installation and on each circuit.

 Safety
If the earth becomes disconnected metal enclosures may reach mains voltage.

Residual current devices

There are devices which are able to detect these very small earth fault currents and automatically disconnect the supply. These are known as Residual Current Devices (RCDs). The internal circuit for one of these is shown in Figure 9.23.

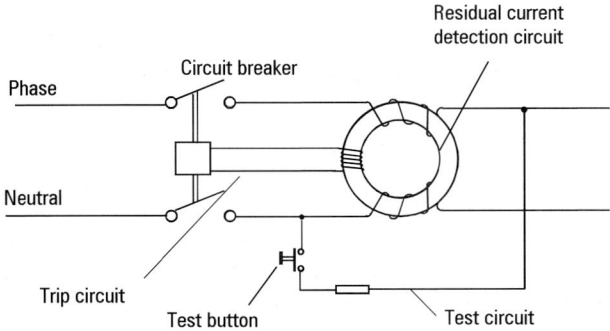

Figure 9.23

They work on the principle of the current in the phase conductor being equal to the return current in the neutral conductor in normal conditions. However, if a phase to earth fault develops the return current in the neutral will be less than that in the phase. This out of balance will be detected in the RCD and it will switch off the supply. These devices can be made so that very small leakages can be detected. The normal setting for sockets in domestic installations is 30 mA.

RCDs can save lives, but as they are an electromechanical device they do develop faults and sometimes without anyone knowing. To ensure they are working correctly they should be tested at regular intervals by pressing the test button. This will check the tripping mechanism in the RCD. It does not, however, test the condition of the circuit it is in. A test with a calibrated RCD tester should be carried out to ensure that the device will trip out at the current and time stated on it.

Main and supplementary equipotential bonding

Within an installation it is important to ensure that all exposed metalwork is at the same electrical potential as that of the consumer's connection to earth. This starts at the intake position, where all gas and water pipes come into the consumer's premises. The metal gas and water pipes must be bonded back to the consumer's main earth terminal (Figure 9.24). Except for PME conditions, the cross-sectional area of these main bonding conductors should not be less than half the main phase conductor and not less than 6 mm^2 or that stated by the local supply authority.

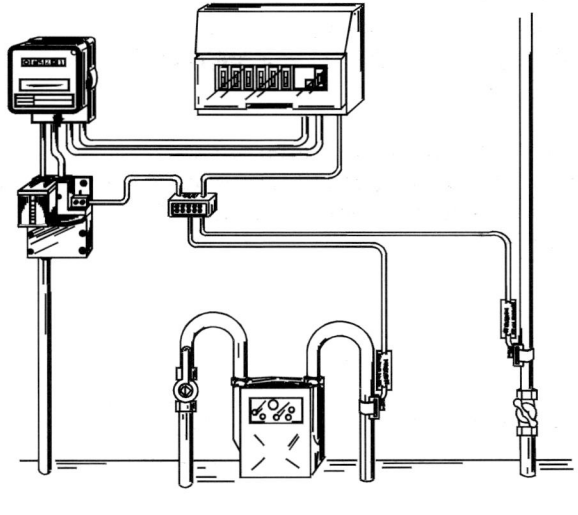

Figure 9.24   Main equipotential bonding conductors at a consumer's intake

In rooms like bathrooms where there are metal fitments and pipework, these must be bonded together including the earth connection of any electrical equipment.

Provided the main and supplementary bonding is correctly carried out and appropriate protective devices are installed, any fault to earth will be cleared without giving rise to danger to persons.

Figure 9.25 shows the metal pipes and exposed metalwork bonded together to create an equipotential area. This means it is not possible for any of these to be at a different potential from another.

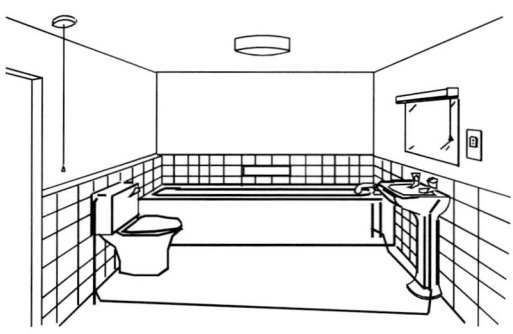

Figure 9.25   Equipotential bonding in bathrooms

Electrical equipment used outside

More and more electrical equipment is being used outside the house. This may be in the garage or garden, where it is not possible to bond everything to the same potential. In these situations it is essential that all electrical equipment is supplied through an RCD so that if a fault develops the supply switches off before any harm can come to the operators.

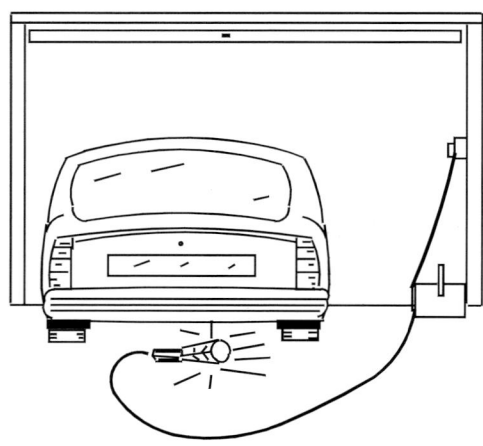

Figure 9.26

Earth electrode test

Where there is a consumer's earth electrode in use this needs testing to ensure that its resistance to earth is not too high. There are special instruments manufactured for this, but it can be carried out using a double-wound transformer, voltmeter and ammeter.

By connecting the circuit in Figure 9.27 to a 12 V isolating transformer it is possible to check that the electrode resistance is acceptable. The auxiliary electrode needs to be placed some distance from the electrode under test, possibly up to 30 metres away. By placing a test probe at regular intervals between the two electrodes and noting the voltage and current readings at each stage, a fairly reliable indication of the earth electrode resistance can be obtained.

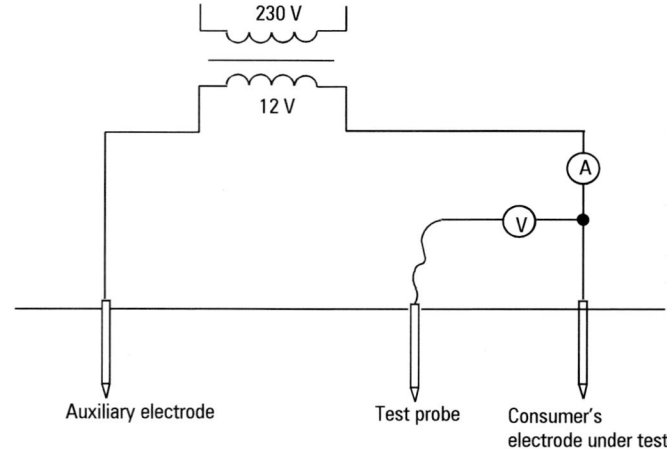

Figure 9.27   Earth electrode test

Points to remember ◀ – – – – – – – – – – – – – –

In general terms the installation must have a connection to _____ and it is important that the _____ path has as _____ a resistance as possible so that in the event of a fault the _____ current can flow, so operating the protection device. An _____ will detect very small earth leakage _____ and disconnect the supply.

Bonding helps to ensure that all parts, electrical and non-electrical, are kept at the same _____ so that the risk of electric shock is kept to a minimum, even under fault conditions.

The _____ _____ _____ impedance of an installation and final circuits must be tested.

Humans are able to perceive very small current flow.

_____ and _____ equipotential bonding needs to be carried out on all electrical installations.

Socket outlets which may be used for equipment _____ must be protected by an RCD.

Draw a typical circuit diagram for an RCD.

Sketch where the main equipotential bonding conductors should be connected for the installation.

Sketch where the supplementary bonding conductors should be connected for the installation.

Self-assessment multi-choice questions

Circle the correct answers in the grid below.

1. The metal pipes are bonded together in a bathroom to
   - (a) use the water pipes as an earth connection
   - (b) make all metalwork at the same electrical potential
   - (c) earth the water supply
   - (d) remind users of the dangers in a bathroom

2. Main bonding conductors connect together gas, water, electricity services and
   - (a) the bathroom fittings
   - (b) the kitchen plumbing
   - (c) the electrical appliances
   - (d) the structural steelwork of a metal-framed building

3. Except where PME conditions apply, the minimum size of bonding conductor should be
   - (a) 2.5 mm^2
   - (b) 4.0 mm^2
   - (c) 6.0 mm^2
   - (d) 10.0 mm^2

4. Portable electrical equipment which is used outside the building must be protected by
   - (a) an isolator
   - (b) a 13 A fuse
   - (c) a 30 mA RCD
   - (d) a 300 mA RCD

5. To ensure that the earth fault path has a sufficiently low impedance, it should be tested by means of
   - (a) an RCD tester
   - (b) an insulation resistance tester
   - (c) continuity tester
   - (d) earth loop tester

Try this

In the course of your work look at the supply intake position of as many different installations as possible and try to decide if the earthing arrangement is:

- (a) provided by means of connection to the supply cable sheath (TN-S)
- (b) provided by means of connection to supplier's neutral (TN-C-S)
- (c) provided by connection to the customer's own earth electrode (TT)

Answer grid

1	a	b	c	d
2	a	b	c	d
3	a	b	c	d
4	a	b	c	d
5	a	b	c	d

10

Inspection and Testing

It could be useful to have available for reference current copies of BS 7671 and/or IEE Guidance Note 3 while working through this chapter.

Complete the following to remind yourself of some important points from the previous chapter.

The main measures for protection listed in the previous chapter are

_____ and _____ are used for protection against overcurrent.

An appliance may be _____ to provide protection against electric shock.

Bonding helps to ensure that all parts of the installation, electrical and non-electrical, are kept at the same voltage so that the risk of _____ _____ is kept to a minimum, even under fault conditions.

On completion of this chapter you should be able to:

◆ describe the procedures that must be undertaken prior to and during inspection and testing of electrical installations
◆ identify the appropriate test equipment for tests of

   continuity of protective conductors

   continuity of ring final circuit conductors

   insulation resistance

   polarity

◆ describe the methods employed when conducting the above tests
◆ list the required sequence of tests
◆ describe the procedures for recording test results
◆ state the requirements for periodic tests

Part 1

In the first book in this series, *Starting Work*, we looked at some basic aspects of inspecting and testing electrical installations. In this chapter we will look at why we inspect and test, what we are looking for when we inspect and how to carry out the tests.

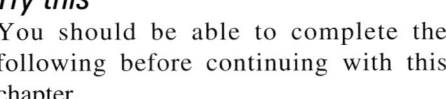 

Try this

You should be able to complete the following before continuing with this chapter.

List the forms of information that should be made available to relevant personnel before they carry out an inspection and test.

Why must the installation, or relevant part, be isolated from the supply before detailed inspection and testing can be commenced?

Why should you liaise with other contractors, or personnel on site, before you commence the inspection?

The aim of an inspection and test on an electrical installation is to verify that

- the installation meets the requirements laid down in the design
- the installation meets all legal requirements and relevant non-statutory regulations
- equipment installed is to the appropriate British Standards
- the installation and equipment is not damaged in any way so as to be dangerous
- the equipment has been installed to the manufacturer's specification and is suitable for the environment in which it is installed

Inspection

Inspections must be undertaken during work on a new installation, at the completion of any new installations, at recommended intervals after completion and when any additions or alterations are made.

Although this is often referred to as a visual inspection it does not preclude the use of other senses. When electrical equipment is overloaded and gets hot it gives off a distinctive smell. This can alert you to possible dangers before any visible signs are apparent. Similarly, loose connections can arc and "crackle".

Remember that as it is necessary to look thoroughly into equipment during the inspection process the installation, or relevant part of it, must be isolated before the inspection commences.

An inspection should include such items as
- the cables
- the cable enclosures, such as conduit and trunking
- all switchgear and control gear
- accessories
- earthing and bonding arrangements
- electrical and mechanical connections

A check list is shown in BS 7671 Regulation 712-01-03.

Documentation

Although many of the checks that have to be carried out will fit into general categories, there are some that will usually require further information. The design documentation should be available throughout the verification process so that it can be referred to as and when it becomes necessary.

Cables

Cables need to be inspected to check whether the correct current rating of cable has been used and to see whether they have been properly installed. The correct fixing distances need to be checked. On older installations cable supports may have come loose or been damaged by mechanical impact or corrosion. Terminations, both of the cable and the conductors,

should be both mechanically and electrically sound. Where cables can be seen they should be visually inspected for damage from heat, corrosion or mechanical impact.

Cable enclosures

Conduit installations need to be checked for many different things. These may depend on whether it is steel or PVC conduit. If it is steel conduit then corrosion can be a problem, so the conduit should be examined to see if the finish of it is suitable for the environment (Figure 10.1). PVC conduit is particularly prone to temperature change and expansion joints should be fitted to allow for it.

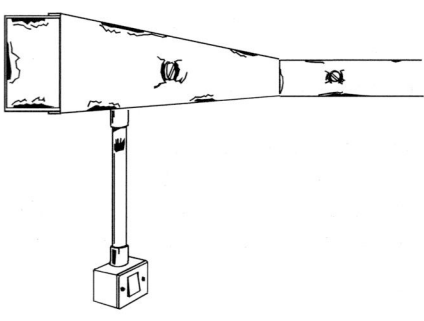

Figure 10.1   Conduit with the incorrect finish for the environment may corrode.

Steel conduit, whether it is used as a circuit protective conductor or not, should be bonded to the electrical earth. PVC and all flexible conduit should contain separate circuit protective conductors. A visual inspection can usually confirm this.

Conduit and trunking, both steel and PVC, should be adequately supported throughout their lengths. As steel trunking has to be electrically continuous, bonding straps should have been fitted across each joint. These may need to be checked to ensure that they are tight. As conduit and trunking are methods of mechanical protection they should be complete with all covers in place and unused holes blanked off (Figures 10.2 and 10.3).

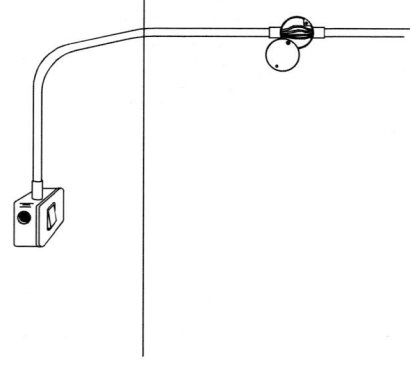

Figure 10.2

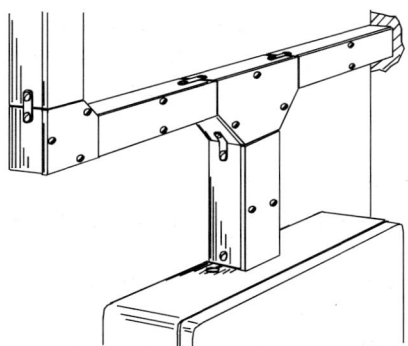

Figure 10.3

An inspection can reveal possible electromagnetic problems, due to phase and neutral conductors being installed through separate entries for example, before any damage can occur.

Where trunking passes through fire barriers such as walls or floors, checks need to be made both inside and outside of the trunking. Outside, the area should be made good to the original specification, and inside a fire barrier should be installed.

> *Remember*
> **Inspection includes using the senses of smell, hearing and touch as well as sight.**
>

Switchgear

Switchgear (Figure 10.4) can be visually checked for damage, corrosion and to see that barriers are in place. However, before protection devices can be checked the documentation stating the rating and type of device must be consulted. It is not only important to know the rating of the device but also the type.

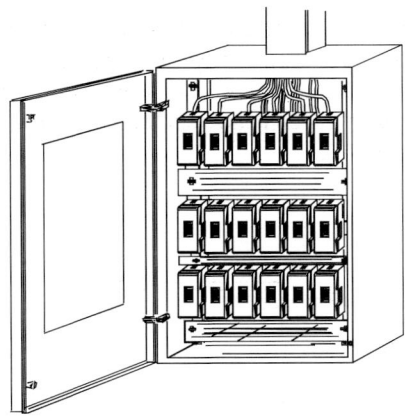

Figure 10.4

The electrical connections need to be checked for tightness. Signs of overheating should always be investigated, as this may indicate loose connections or overloaded cables.

Accessories

Accessories can be a source of loose connections, as often a number of conductors have to go into one terminal. These need to be checked for poor terminations (Figure 10.5) and for signs of possible overheating.

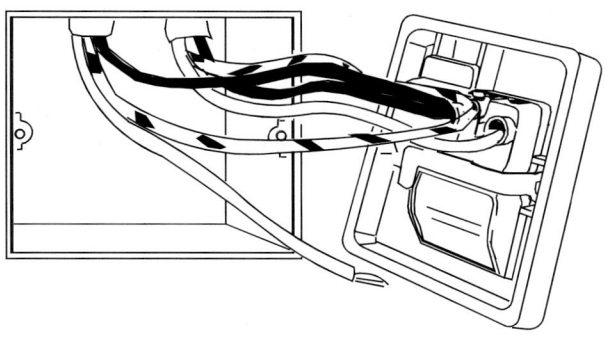

Figure 10.5

> *Remember*
> **An inspection verifies that the workmanship and equipment is installed to the specified standards.**
>

Earthing and bonding

Earthing at the main intake should be checked to see if it is complete and correctly labelled. If an earth electrode is installed, as in a TT system (Figure 10.6), all connections and identification labels should be checked. Each circuit protective conductor should be identified by colour coding as they go to each circuit. These need to be inspected to ensure bare conductors are sleeved.

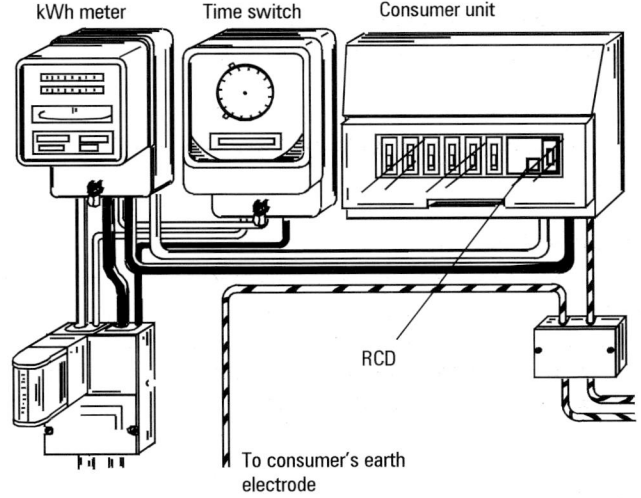

Figure 10.6   *TT system*

Main equipotential bonding, to ensure an equipotential zone, should be carried out between all services. A visual inspection may need to be made to examine connections to ductwork, heating pipes, structural steel and so on. This inspection should include checks on the use of the correct size, identification and labelling of conductors.

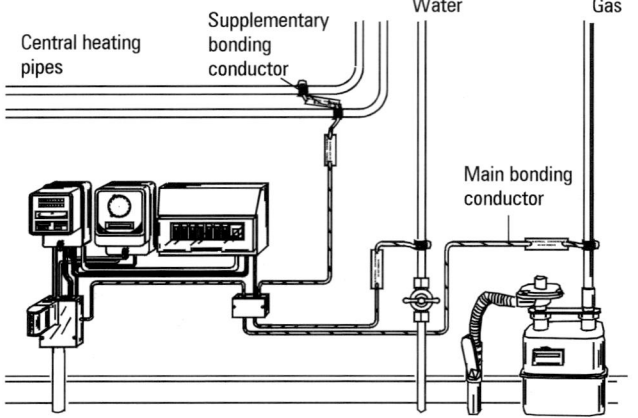

Figure 10.7   Visual inspections should be carried out to confirm that connections are satisfactory.

Supplementary bonding (Figure 10.7) is used to complete the equipotential zone, and checks need to be made for good connections and correct identification.

Residual Current Devices (RCDs) are used in installations as an added precaution. Checks need to be made to ensure that the correct rating and time setting is used. When the supply is connected the test button should be pressed to check that the mechanical parts of the RCD are still in working condition.

Type of circuit
Inspections also have to relate to the type of circuits used. These include lighting circuits, ring circuits and bathrooms The labelling of circuits must be checked to ensure that it is correct.

Lighting circuits need to be checked to ensure that the correct rating of protection device has been installed. Luminaires should be fitted correctly with a suitable size and type of flexible cable. Checks should be made to ensure that the correct lamps are fitted to luminaires and the rating has not been exceeded. Where necessary, heat resisting sleeving or cable should be used (Figure 10.8).

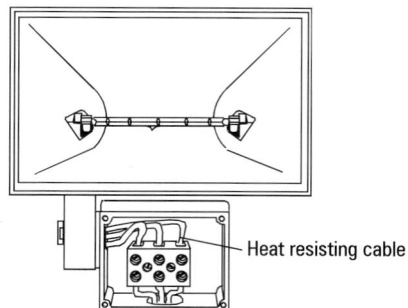

Heat resisting cable

Figure 10.8   Where high temperatures are experienced heat resisting cable must be used.

Consideration should be given to the positions of switches and the use of cord grips in ceiling roses and lampholders.

A note may need to be made of where electronic equipment is in use so that it can be isolated before tests are carried out.

Ring circuits with socket outlets to BS 1363 have features in common with other circuits; these include the correct rating of protection devices and cable sizes. But they are different because most other circuits are radials. It is therefore important to ensure that a ring circuit is wired correctly. In some circumstances a visual inspection will show this prior to testing.

Bathrooms (Figure 10.9), by their very nature, are a potential danger area for the use of electrical equipment. This means that special consideration has to be given to the correct positioning of electrical equipment and accessories. If there are isolated sections of pipework a complete bonding system may be necessary. The special Regulations in BS 7671, Section 601, will need to be given consideration.

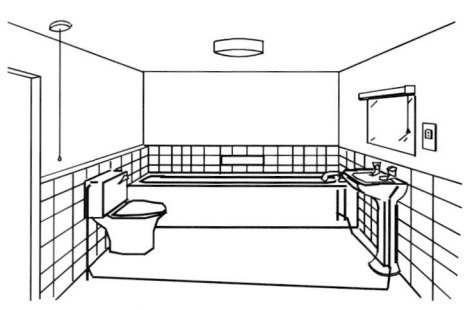

Figure 10.9   Equipotential bonding cables and connections should be checked.

Warning notices
There are several occasions when warning notices of one type or another should be installed (Figure 10.10). Checks should always be made to see that they are in place and that they meet the requirements of the appropriate Regulations.

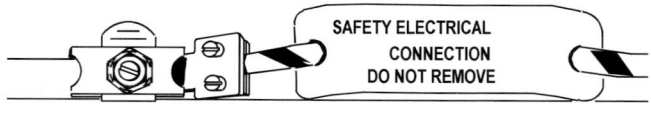

SAFETY ELECTRICAL
CONNECTION
DO NOT REMOVE

Figure 10.10

⚠️ Inspections usually mean taking the covers off equipment and checking inside.
All necessary safety precautions need to be taken when this is carried out.

BS 7671:1992 provides guidance on items included in the inspection process.

The process includes the use of the senses smell, touch, hearing and sight.

Suitability of equipment for the environment should be considered together with condition.

Enclosures should be complete and all unused entries should have been closed.

Inspection includes the condition of cables, insulation, terminations and connections.

The inspection is carried out against current requirements, as detailed in BS 7671, irrespective of when the installation was installed.

Try this

1. Explain the main purposes of carrying out inspections and tests on any electrical installation.

2. Explain why old installations that have to be inspected and tested have to be approached differently from new installations.

Part 2

The tests

The tests we will be describing are:
- continuity of protective conductors
- continuity of ring final circuit conductors
- insulation resistance
- polarity

These tests are carried out before the supply is connected, in the case of new buildings, or isolated in the case of installations that have been altered. An example of an isolation procedure flowchart is given below.

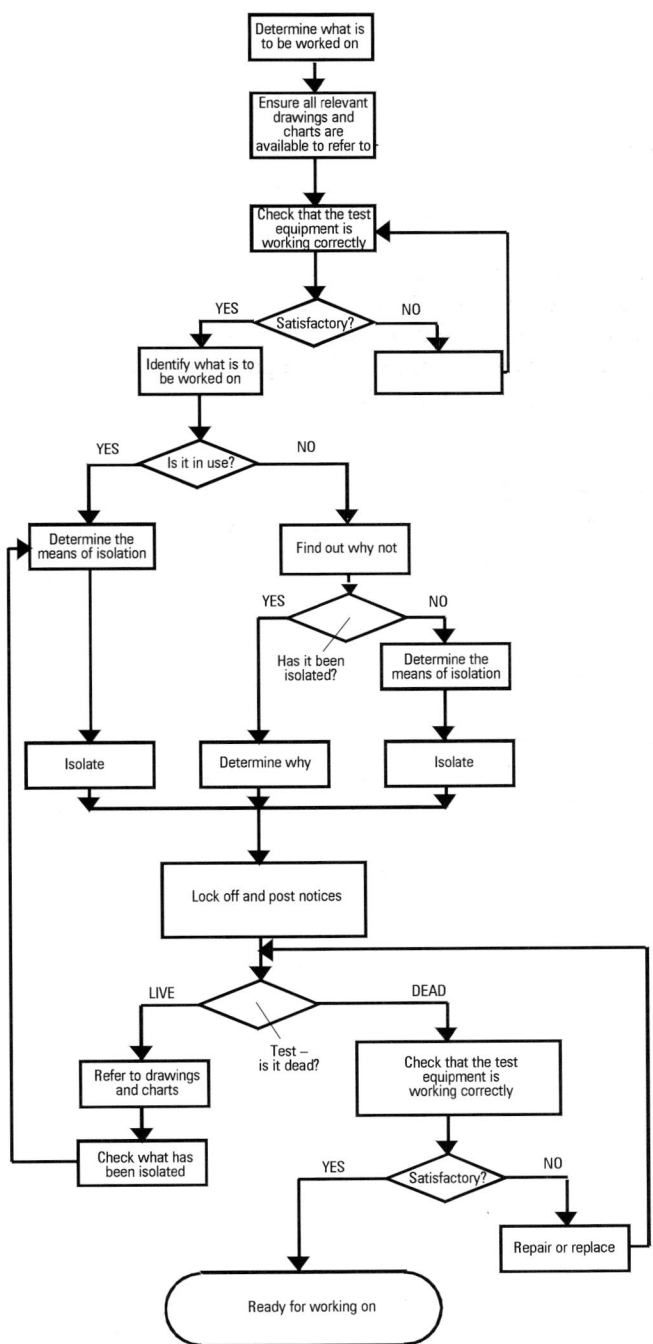

Test instruments

The test equipment required for the tests being considered are as follows:

Tests	Test equipment
Continuity of protective conductors	Low-resistance ohmmeter (milliohmmeter)
Continuity of ring final circuit conductors	Low-resistance ohmmeter (milliohmmeter)
Insulation resistance	High-resistance megohmmeter with d.c. test voltages to suit the installation under test, i.e. 250 V, 500 V or 1000 V when loaded with 1 mA. Resistance range needs to be in excess of 1 $M\Omega$
Polarity	Low-resistance ohmmeter (milliohmmeter)

To ensure that the readings obtained in tests are as accurate as possible, the test instruments should be continually checked. In some cases this is a case of zeroing the scale with leads shorted out. In others, battery checks may have to be carried out.

It is important that readings are accurately transferred from the test equipment and recorded, and where multirange instruments are used care has to be taken to ensure that the correct scale and range are read.

 Test probes used for confirming isolation must conform to the guidance in the Health and Safety Executive "Electrical test equipment for use by electricians". When carrying out tests it is the tester's responsibility to observe all safety procedures to ensure his or her own safety and that of others.

Continuity of protective conductor test

The first test that must be carried out is to confirm that the circuit protective conductors are continuous throughout the circuit and that all connections are sound. There is more than one way to carry out these tests and it will depend on the particular circumstances as to which will be the most suitable.

Method 1, the $R_1 + R_2$ method, is useful to assess the continuity of circuit protective conductors and their associated phase conductors. Method 2, the long lead method, is more appropriate for testing the continuity of main and supplementary bonding conductors. For both methods the supply must be securely isolated and all necessary documentation must be available before starting the test. Documentation should consist of the circuits under test and the design resistance values. The test results should be less than the design values.

The test equipment required is an ohmmeter with a low resistance scale, capable of measuring values in milliohms.

Test method 1

After securely isolating the supply disconnect any supplementary bonds to the protective conductors. Make a temporary link at the supply end between the phase and the cpc of the circuit under test (Figure 10.11). Now measure the resistance between the phase and cpc at each point, such as light fittings, switches and so on, in the circuit. The value obtained is known as the $R_1 + R_2$ value, and the highest measured value for each circuit should be recorded and compared with the design value. If the reading is greater than the design value it indicates that there is something wrong that needs to be investigated.

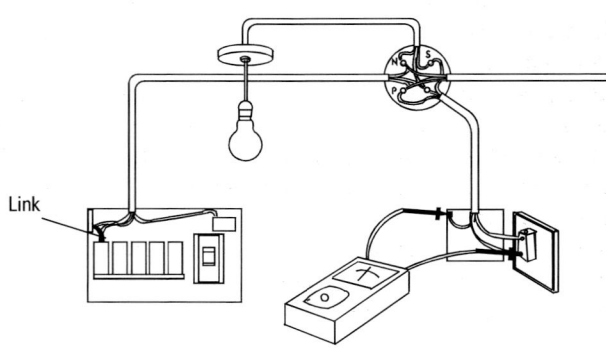

Figure 10.11   Test method 1 ($R_1 + R_2$)

After completing the test remove the temporary link at the supply end of the circuit and reconnect any bonding conductors that were removed. The $R_1 + R_2$ method also verifies polarity.

Test method 2

Securely isolate the supply. From the consumer's main earth terminal attach a long test lead to the ohmmeter. With the other ohmmeter lead make contact with the protective conductors at points on the circuit (Figure 10.12). In this method, as the distance between testing points can be quite long, the resistance of the test leads must be taken and this reading subtracted from each test result. For example, if a test resistance is 0.7 Ω and the test leads have a resistance of 0.2 Ω, then the actual resistance of the test is 0.7–0.2 = 0.5 Ω.

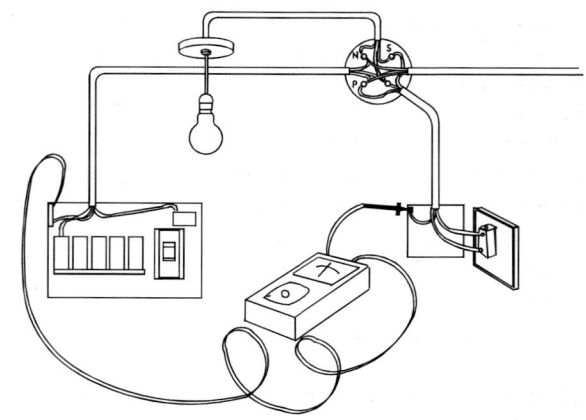

Figure 10.12   Test method 2 (long lead)

Alternatively, some meters can be zeroed with the long leads connected.

Record the values taken and deduct the values of the test leads as above. The resistance of the protective conductor R₂ should be recorded on the appropriate form.

Continuity of ring final circuit conductors

The ring final circuit starts and finishes at the same point and, as the name implies, should form a continuous ring. The phase starts and finishes at the same protective device (fuse or circuit breaker), the neutral starts and finishes at the same connection on the neutral bar, and the circuit protective conductor starts and finishes at the same connection on the earth block. Each socket outlet on the ring has at least two cables to it. This means that there must be at least two conductors in each terminal. The red conductors are connected to the L (phase terminal), black to N (neutral) and the bare circuit protective conductors green/yellow sleeving to the E (earth) terminal.

The continuity of ring final circuit conductors is a test to verify that the ring is complete and has not been interconnected. Ring circuits, if they are not connected correctly, can become a fire hazard.

There are several methods of carrying out this test, and at this stage we will only be looking at the one detailed in IEE Guidance Note 3, Inspection and Testing.

The tests are carried out before the supply is connected using a low reading ohmmeter. The first test is to confirm that there is a circuit between the two ends of the ring circuit cables. In this circuit there are three separate rings, one for the phase conductors, one for the neutral and one for the circuit protective conductors. If the protective circuit had been in steel conduit or trunking this would not have to be connected in a ring.

To verify the continuity of the ring circuits they are first disconnected at the distribution board and brought out in their pairs. The resistances of the conductors are taken and recorded, phase to phase, neutral to neutral and cpc to cpc. This measures the resistance, end to end, of each conductor. The values for the phase and neutral conductors should be about the same value. Generally the cpc is a reduced cross-sectional area (for example 2.5 mm² phase and neutral with 1.5 mm² cpc), so the value for the cpc should be higher, around 1.7 times that of the phase. This test confirms that the circuit is a ring and the correct conductors have been identified. If appropriate values are not obtained further investigation is required.

The phase conductor of one pair should then be connected to the neutral of the other and the remaining phase and neutral conductors should be similarly connected. The resistance is measured between the cross-connected pairs (Figure 10.13).

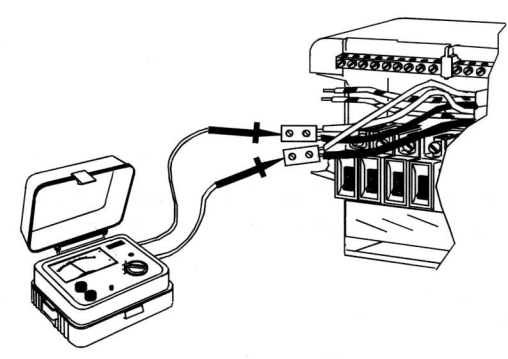

Figure 10.13   Ring circuit continuity test

The reading obtained should be approximately half the values we obtained measuring phase to phase or neutral to neutral. With the phase and neutral conductors still "cross-connected" a reading should be taken at each socket on the ring. The values obtained should be substantially the same as those we had at the distribution board. Remember that the resistance at outlets spurred from the ring will be higher, this value depending upon the length of cable in the spur.

We now cross-connect the phase and cpc conductors in the same way (Figure 10.14) and take a reading at the distribution board. This value should be approximately

$$\frac{\text{phase to phase } \Omega + \text{cpc to cpc } \Omega}{4}$$

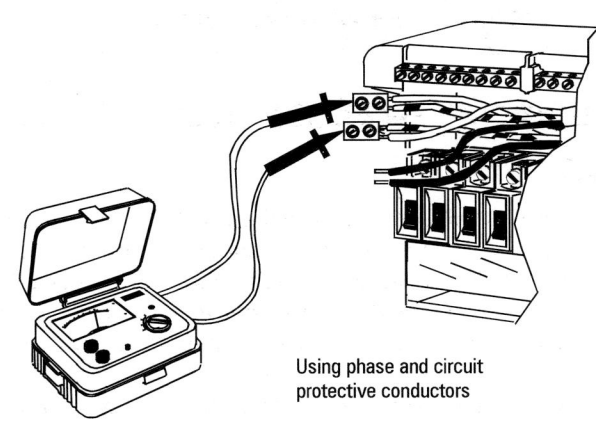

Using phase and circuit protective conductors

Figure 10.14   Ring circuit continuity test

A reading is then taken at each socket outlet, which, as before should be the same as that obtained at the distribution board.

This will ensure that the conductors are connected as a ring circuit and confirm polarity at each socket.

Insulation resistance

 This test requires voltages of up to 500 V d.c. Care must be taken to avoid accidents.

The first two tests we have looked at were to prove that there was a circuit and that the resistance was low enough.

The insulation resistance test is to confirm that the insulation throughout the circuit has not been damaged.

Before the test is carried out it is important to check that
* there is no supply to the circuit being tested
* all lamps are removed
* all equipment that would normally be in use is disconnected
* any electronic equipment that would be damaged by the high-voltage test is disconnected (this may include lamp dimmer switches, delay timers, RCDs and so on)
* all fuses are in place
* all switches are in the ON position (unless they protect equipment that cannot otherwise be disconnected)

For installations supplied at voltages up to 500 V a.c. the test voltage is 500 V d.c.

The test must be carried out between phase and neutral, phase and earth, and neutral and earth.

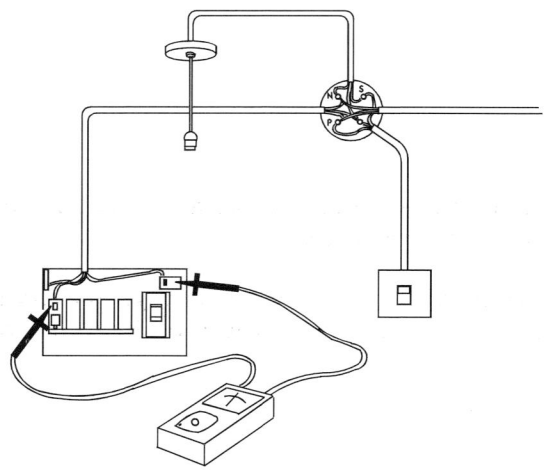

Figure 10.15   Insulation resistance test

Each reading should be at least 0.5 MΩ on a circuit such as those included in this section to comply with BS 7671. However where an insulation resistance is recorded at less than 2 megohm there is a possibility of a latent defect. The reading may indicate INF or ∞ (infinity). This means the reading is greater than the instrument can indicate. Values should not be recorded as ∞ or INF but as "greater than" the highest reading on the instrument scale. So if, for instance, the instrument maximum reading is 50 MΩ and the result is higher than that at ∞, we should record > 50 MΩ where ">" indicates "greater than".

Polarity

A polarity test checks that the installation connections have been made in the correct conductors. Much of the polarity test can be done when doing the $R_1 + R_2$ test method for the continuity of protective conductors.

On domestic installations the tests should verify that:
* all lighting switches are connected in the phase conductor
* the centre pin of any ES (Edison Screw) lampholders are connected in the phase conductor
* the switches on socket outlets are connected in the phase conductor
* the correct pin of socket outlets is connected to the phase conductor
* where double pole switches are used, such as for immersion heaters, the phase and neutrals have not been swapped over

A visual inspection should have confirmed the correct identification of the conductors so this should be assumed correct unless proved otherwise. This can be carried out with a meter or a bell tester.

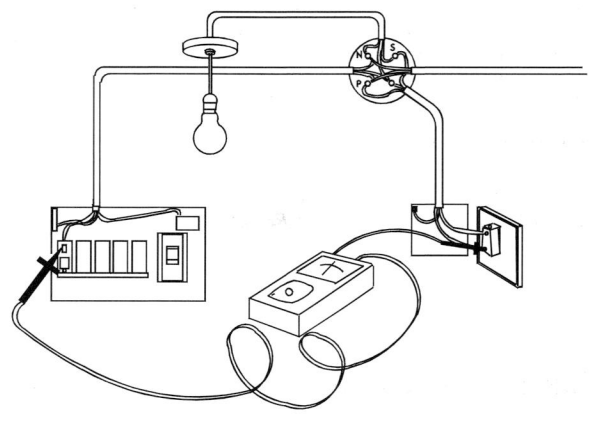

Figure 10.16   Polarity test

Remember
The polarity of all equipment must be confirmed *before* the installation is energised. Plug-in testers which require a supply may only be used as a check once the supply is connected.

The test results

When the inspection and test have been completed, the results must be recorded on the form prescribed in BS 7671 Appendix 6 and, together with a schedule of test results, forwarded to the person ordering the report.

Example of schedule of test results

Installation:	Distribution Board Ref:	Instruments used:			
Contractor:	Earthing Arrangements:			**Make**	**Serial No.**
Engineer:	External Impedance: Z_e	Continuity			
Date of Tests:	Maximum Prospective Short Circuit Current: I_f DB	Insulation			
	Supply Voltage:	Loop Impedance			
		RCD			

Description of work completed *Rewire office 3 phase dis board*

Distribution Board	No. of points	Fuse BS(EN) type	Rating (A)	Cable size mm^2	length m	cpc mm^2	Test Results					
							Z_s Ω	Ins Res MΩ	Polarity ✓	RCD mS	Ring Continuity ✓	$R_1 + R_2$
										1 × | 5 ×		
1. Lights	6	21361	5	1.5	15	1.0	0.5	10		- |	-	

Figure 10.18   A Record of Test Results

Points to remember ◀ – – – – – – – – – – – – – –

Do not attempt to carry out any inspection or testing work unless you are competent to do so. When carrying out tests it is important to follow the _____ procedure and ensure that it is safe to proceed.

Test instruments, if incorrectly applied, will produce false results. This could create an extremely dangerous situation for persons or livestock coming into contact with a faulty installation.

Record the make, model and _____ _____ of instruments used to carry out the tests.

Ensure the correct instrument is used for each test, with the appropriate scale selected and reading taken in each case.

Carry out testing in the correct sequence.

Remember to subtract the resistance of any leads used for testing, where there is no facility to zero the instrument with the leads connected.

Record the values obtained for each test and compare these with the design maximum.

The testing must not give rise to damage to equipment or danger to persons, livestock and property during its execution.

Remember
The tests, and the order in which they should be undertaken, are:

1. **continuity of protective conductors, main and supplementary bonding**
2. **continuity of ring final circuit conductors**
3. **insulation resistance**
4. **polarity**

Self-assessment multi-choice questions
Circle the correct answers in the grid below.

1. Which of the following is not included in the aim of an inspection and test on an electrical installation
   (a) condition of wiring
   (b) correct operation of appliances
   (c) correct operation of functional switches
   (d) compliance with appropriate standards
2. Place the following tests in their correct sequence
   1. insulation resistance
   2. polarity
   3. cpc continuity
   4. ring circuit continuity
   (a) 3 4 1 2
   (b) 3 2 1 4
   (c) 4 2 1 3
   (d) 1 2 3 4
3. Continuity of circuit protective conductors is measured using a
   (a) polarity tester
   (b) insulation resistance tester
   (c) milliohmmeter
   (d) earth loop tester
4. An insulation resistance test on a 400 V installation should be carried out at
   (a) 250 V
   (b) 400 V
   (c) 500 V
   (d) 800 V
5. Correct polarity must be verified
   (a) by a visual check only
   (b) during construction
   (c) once supply is available and the installation is energised
   (d) at the distribution board only

Answer grid
1	a	b	c	d
2	a	b	c	d
3	a	b	c	d
4	a	b	c	d
5	a	b	c	d

End test

Circle the correct answers in the grid at the end of the multi-choice questions.

1. For electricity supply purposes, the standard unit of energy supplied is the
   (a) watt
   (b) kilowatt
   (c) kilowatt hour
   (d) kilowatt per hour

2. A domestic tariff which charges less than half the daytime rate for energy consumed between midnight and 7 a.m. is called
   (a) Economy 7
   (b) Standard tariff
   (c) Off-peak
   (d) Night rate

3. A domestic consumer with an average consumption of 10 units per day at a rate of 8.42 pence per unit and a standing charge of £11.67 could expect to receive a quarterly bill (excluding VAT) of
   (a) £76.62
   (b) £88.29
   (c) £104.88
   (d) £7673.87

4. As a conductive material, copper is only surpassed by
   (a) iron
   (b) brass
   (c) aluminium
   (d) silver

5. One advantage which aluminium possesses over copper is that it is
   (a) lighter
   (b) more flexible
   (c) stronger
   (d) a better conductor

6. The insulation of overhead conductors is normally provided by
   (a) a PVC sheath
   (b) a rubber covering
   (c) a ceramic sleeve
   (d) the surrounding air

7. Aluminium overhead conductors may be strengthened by incorporating a core made of
   (a) copper
   (b) steel
   (c) brass
   (d) zinc

8. The purpose of the PVC outer sheath on a PVC-insulated and sheathed cable is to
   (a) act as an additional insulator
   (b) help to identify the cable type
   (c) provide mechanical protection
   (d) prevent heat loss

9. A single, circular-section conductor with a diameter of 2.256 mm has a cross-sectional area of
   (a) 4 mm^2
   (b) 7.08 mm^2
   (c) 9.87 mm^2
   (d) 16 mm^2

10. The mineral insulating medium used in mineral insulated cable is
    (a) magnesium oxide
    (b) polyvinyl chloride
    (c) calcium chloride
    (d) manganese dioxide

11. The standard sizes of conduit available include
    (a) 15 mm
    (b) 25 mm
    (c) 35 mm
    (d) 40 mm

12. A type of conduit finish which is resistant to corrosion and may be found in damp situations is known as
    (a) electro-plated
    (b) black enamelled
    (c) sherardized
    (d) shermanized

13. To avoid the effects of current induced in a steel conduit system, which one of the following practices must be observed
    (a) phase and neutral conductors must be run in separate steel conduits
    (b) three-phase conductors may be kept together but the neutral run separately
    (c) each phase conductor and the neutral (if any) must be enclosed separately
    (d) all phase conductors and the neutral of any circuit must be run in the same metallic enclosure

14. Using the appropriate table from Guidance Note 1 select the maximum spacing between supports for a horizontal run of 20 mm steel conduit
    (a) 1.5 m
    (b) 1.75 m
    (c) 2.0 m
    (d) 2.25 m

15. The same table will give the maximum distance between supports for a vertical 25 mm heavy gauge rigid insulating conduit as
    (a) 1.5 m
    (b) 1.75 m
    (c) 2.0 m
    (d) 2.25 m

16. Where trunking passes through a wall or partition, which one of the following steps may be necessary
    (a) a separate cpc must be installed
    (b) the cable sizes must be completely recalculated
    (c) all conductors must be terminated in a junction box
    (d) the inside of the trunking must be sealed to prevent the spread of fire

17. The minimum distance between supports for cables installed in a vertical run of trunking is
    (a) 3 m
    (b) 4 m
    (c) 5 m
    (d) 6 m

18. "Dado" trunking is a name used to describe a type of trunking which is mounted on the wall at
    (a) skirting level
    (b) 300 mm above floor level
    (c) just above "desktop" level
    (d) just below the ceiling

19. The type of trunking which may be incorporated as part of a door frame assembly is described as
    (a) architrave trunking
    (b) door trunking
    (c) mini-trunking
    (d) frame trunking

20. Busbar trunking is frequently found in factory installations because
    (a) it is easy to install
    (b) it has an attractive appearance
    (c) there is the facility for making tap-off connections which are close to load positions
    (d) its installation costs are low

21. Where non-standard cables are used, the space occupied by cables should not exceed what percentage of the available space
    (a) 35%
    (b) 45%
    (c) 55%
    (d) 65%

22. Up to how many cables of 3.57 mm diameter may be installed in a trunking measuring 60 mm × 30 mm
    (a) 41
    (b) 50
    (c) 81
    (d) 504

23. Using the tables provided in Guidance Note 1 determine how many 6 mm^2 cables may be added to 24, 4.0 mm^2 cables already installed in a 50 mm × 50 mm trunking to fill it to its maximum capacity and select the nearest number.
    (a) 30
    (b) 45
    (c) 49
    (d) 54

24. In a short straight run of 25 mm conduit 6 × 1.5 mm^2 solid conductors are already installed. Which of the following numbers indicates the greater number of 1.0 mm^2 singles that may be added before exceeding the total permitted?
    (a) 9
    (b) 15
    (c) 24
    (d) 29

25. According to the information given in IEE Guidance Note 1, if a 32 mm conduit is given a factor of 900, a comparable 50 mm conduit would have a factor of
    (a) 1406
    (b) 1800
    (c) 2340
    (d) 2500

26. The conduit factors which help to determine the capacity of a conduit are determined by the diameter, the length of run and
    (a) the size of the cable to be installed
    (b) the type of the cable to be installed
    (c) the ambient temperature
    (d) the number of bends

27. Perforated cable tray is a support system for cables which is frequently found in
    (a) domestic installations
    (b) shops and office premises
    (c) hotels and restaurants
    (d) industrial premises

28. Cables which are fixed to tray in a single layer without spaces between them are
    (a) rated at their full current-carrying capacity
    (b) permitted to carry more than their full capacity because of the improved heat dissipation
    (c) de-rated by the appropriate grouping factor
    (d) not permitted by BS 7671

29. In addition to that manufactured from mild steel, cable tray is also made from
    (a) copper
    (b) brass
    (c) tin plate
    (d) stainless steel

30. Where bell-pushes are connected in series as in a "closed circuit" system, their contacts must be
    (a) normally closed
    (b) normally open
    (c) change-over
    (d) double-pole

31. A "maintained" emergency lighting system is
    (a) designed to operate constantly with or without mains power
    (b) subjected to a rigorous maintenance programme
    (c) designed to come into operation when the mains voltage fails
    (d) serviced by the manufacturer every three months

32. A switching system which controls one or more lights from three locations is called
    (a) one-way
    (b) two-way
    (c) three-way
    (d) intermediate

33. (i) Light is produced in an incandescent lamp by passing current through a conducting filament.
    (ii) Light is produced in a discharge lamp by passing current through a gas or vapour.
    (a) only statement (i) is true
    (b) only statement (ii) is true
    (c) both statements are true
    (d) both statements are false

34. The assumed current demand of a 45 A domestic cooker is
    (a) 20.5 A
    (b) 22.5 A
    (c) 45 A
    (d) 49.5 A

35. Which of the following is recognised as a "special installation or location" under Part 6 of BS 7671?
    (a) bathroom or shower room
    (b) petrol filling station
    (c) underground car park
    (d) high-voltage substation

36. Miniature circuit breakers to BS EN 60898 are available in ratings of 6 A, 10 A, 16 A and also
    (a) 24 A
    (b) 25 A
    (c) 30 A
    (d) 32 A

37. Where a connection to earth is not provided by the electricity supply company, the consumer must provide an earth connection to
    (a) the water supply pipe
    (b) the gas supply pipe
    (c) an earth electrode
    (d) the incoming neutral

38. It is recommended that the inspection of an installation should be carried out
    (a) before the testing procedure
    (b) after the testing procedure
    (c) during the testing procedure
    (d) within three months of the completion of the work

39. The continuity test is used to determine the resistance of
    (a) the cable insulation
    (b) the supply conductors
    (c) the neutral conductors
    (d) the circuit protective conductors

40. A polarity test is used to ensure that all fuses and single-pole switches are in the phase conductor only and
    (a) all circuits are correctly identified
    (b) all fuses are correctly rated
    (c) there are no short circuits in the installation
    (d) the centre contact of all E.S. lampholders is connected to the phase side of the supply.

Answer grid

1	a	b	c	d	21	a	b	c	d
2	a	b	c	d	22	a	b	c	d
3	a	b	c	d	23	a	b	c	d
4	a	b	c	d	24	a	b	c	d
5	a	b	c	d	25	a	b	c	d
6	a	b	c	d	26	a	b	c	d
7	a	b	c	d	27	a	b	c	d
8	a	b	c	d	28	a	b	c	d
9	a	b	c	d	29	a	b	c	d
10	a	b	c	d	30	a	b	c	d
11	a	b	c	d	31	a	b	c	d
12	a	b	c	d	32	a	b	c	d
13	a	b	c	d	33	a	b	c	d
14	a	b	c	d	34	a	b	c	d
15	a	b	c	d	35	a	b	c	d
16	a	b	c	d	36	a	b	c	d
17	a	b	c	d	37	a	b	c	d
18	a	b	c	d	38	a	b	c	d
19	a	b	c	d	39	a	b	c	d
20	a	b	c	d	40	a	b	c	d

Answers

These answers are given for guidance and are not necessarily the only possible solutions.

Chapter 1

p.1 230 V, 400/230 V, 11 kV, 110 V a.c.; Off the floor away from the damp; Boxed in their wrapping to prevent discolouring due to dirt and sunlight.

p.3 Try this: 2591.1

p.4 Try this: £64.82; Try this: £54.83

p.5 Try this: £81.87

p.6 SAQ (1) b; (2) b; (3) d; (4) c; (5) a

Chapter 2

p.15 SAQ (1) d; (2) c; (3) b; (4) a; (5) a; (6) d; (7) b; (8) c; (9) d; (10) c

p.17 Try this: 25 mm^2; 47.2 mm^2 (45 mm^2); 147 mm^2 (150 mm^2); 389 mm^2 (400 mm^2)

p.19 Try this: To make the cable more compact; 35 mm^2; 2.16 mm; the statement (b) is correct

p.21 PVC sheath; PVC sheath and aluminium tape screen; copper (or aluminium) sheath

p.24 SAQ (1) d; (2) c; (3) b; (4) b; (5) c; (6) d; (7) c; (8) a; (9) d; (10) b

Chapter 3

p.30 SAQ (1) c; (2) d; (3) c; (4) a; (5) c

p.32 Try this: (1) 25 A; (2) 4.35 A

p.37 Try this: (1) 50; (2) 0.8; (3) 52.17; (4) 1

p.37 SAQ (1) c; (2) c; (3) d; (4) a; (5) d

p.39 Try this:

240	9.6
110	4.4
200	8
220	8.8
250	10
415	16.6

p.42 SAQ (1) b; (2) c; (3) b; (4) b; (5) c

Chapter 4

p.48 SAQ (1) d; (2) a; (3) c; (4) a; (5) b; (6) b; (7) a; (8) a; (9) c; (10) c

Chapter 5

p.52 Try this: (1) welded seam galvanised conduit; (2) seamless solid drawn conduit; (3) welded seam black enamel conduit

p.54 Try this: (1) 40 mm; (2) 62.5 mm; (3) 80 mm

p.55 SAQ (1) c; (2) a; (3) b; (4) c; (5) c

p.59 SAQ (1) d; (2) b; (3) c; (4) c; (5) b

p.64 Try this: (1) ✗; (2) ✔; (3) ✔; (4) ✔; (5) ✔; (6) ✔; (7) ✔; (8) ✗; (9) ✗; (10) ✗

p.66 SAQ (1) a; (2) c; (3) b; (4) c; (5) b; (6) a; (7) c; (8) b; (9) c; (10) c

p.72 SAQ (1) d; (2) b; (3) d; (4) d; (5) c; (6) d; (7) a; (8) c; (9) a; (10) c

pp.73 and 74 Progress check

(1) c; (2) d; (3) d; (4) c; (5) a; (6) b; (7) d; (8) d; (9) a; (10) b; (11) d; (12) c; (13) d; (14) d; (15) a; (16) c; (17) c; (18) d; (19) b; (20) d; (21) b; (22) a; (23) c; (24) a; (25) d

Chapter 6

p.75 3 m; 38 mm × 33 mm to 225 mm × 100 mm

p.80 SAQ (1) c; (2) b; (3) d; (4) d; (5) a

p.86 SAQ (1) d; (2) b; (3) b; (4) c; (5) c

p.88 Try this: fifty 6 mm^2 cables

p.89 Trunking size 100 mm × 25 mm

p.91 Try this: 15 cables

p.91 Try this: No, 14 cables are the maximum

p.92 SAQ (1) a; (2) b; (3) c; (4) c; (5) b

Chapter 7

p.96 SAQ (1) c; (2) a; (3) c; (4) b; (5) d

p.97 Try this: 150 mm

p.100 Try this: (1) 100 mm; (2) (a) 300 mm, (b) 400 mm

p.100 SAQ (1) a; (2) d; (3) c; (4) d; (5) d

Chapter 8

p.106 Try this: (1) 2.99 kW; (2) 6.9 kW

p.106 SAQ (1) b; (2) c; (3) c; (4) d; (5) a

p.108 Try this: 4600 watts

p.108 Try this: 30.47 A

p.112 SAQ (1) c; (2) a; (3) b; (4) b; (5) c; (6) c; (7) c; (8) b; (9) c; (10) b

Chapter 9

p.119 SAQ (1) b; (2) d; (3) d; (4) b; (5) c; (6) a; (7) c; (8) c; (9) b; (10) c

p.124 SAQ (1) b; (2) d; (3) c; (4) c; (5) d

Chapter 10

p.125 isolation; automatic protection against overload; automatic disconnection in the event of current leakage to earth

p.134 SAQ (1) b; (2) a; (3) c; (4) c; (5) b

pp.135, 136 and 136: End test

(1) c; (2) a; (3) b; (4) d; (5) a; (6) d; (7) b; (8) c; (9) a;
(10) a; (11) b; (12) c; (13) d; (14) b; (15) b; (16) d;
(17) c; (18) c; (19) a; (20) c; (21) b; (22) c; (23) a; (24) d;
(25) c; (26) d; (27) d; (28) c; (29) d; (30) a; (31) a; (32) d;
(33) c; (34) a; (35) a; (36) d; (37) c; (38) a; (39) d; (40) d

Appendix

Terms used in this book

alarm systems
alarm and emergency circuits can be "open circuit systems" where "push to make pushes" or other detection devices may be used to set off the alarm or "closed circuit systems" where the whole detection circuit is complete until something breaks it and sets off the alarm. (pp. 101 and 102)

ambient temperature
the temperature of the surroundings (p. 34)

architrave trunking
trunking around doorways and the like (p. 81)

armoured cable
a cable with a layer of metallic armouring to protect it from damage due to impact (p. 11)

bench trunking
trunking around benching in laboratories for example (p. 82)

bending machine
a machine to put a bend in conduit (p. 51)

bending radius
so that damage is not caused to a cable there is a minimum bending radius that should be taken into account. This also applies to the conduit, trunking or cable tray that the cable is installed in or on (p. 27)

bending spring
a bending spring is used to make bends in PVC conduit (p. 61)

brackets
can be used to support trunking depending on the location (p. 79)

busbar trunking
a specialist trunking which does not contain cables but has copper or aluminium busbars which are continuous throughout the length of the trunking system (p. 84)

bush spanner
a bush spanner helps to tighten the bush when conduit is terminated into boxes so that continuity is maintained (p. 52)

cable
a means of carrying current (p. 7)

cable tray
a support system for sheathed cables used in industrial/commercial installations (p. 93)

catenary wire
a galvanised steel catenary wire carries the weight of the cable where the cable is suspended between two buildings (p. 26)

channel
channel provides protection for sheathed cable commonly used in carcass wiring of domestic installations (p. 70)

circuit breaker
a device that will automatically disconnect a circuit from the supply in the event of an overload (p. 115)

circuit protective conductor
a conductor used for protection against electric shock (p. 45)

clamp
clamps can be used with clips to fit conduits to roofs comprising a steel girder construction (p. 57)

clips
see clamp (p. 57)

compartmentalised trunking
trunking that is separated into compartments to keep circuits in voltage Band I and voltage Band II apart (p. 81)

conductor
the electrical conductor is the part of the cable that carries the current (p. 7)

conduit
basically an enclosure in the shape of a pipe which provides mechanical protection for single insulated cables (pp. 44 and 49)

conduit boxes
as there is a limit to the length of conduit through which we can pull cables in one go conduit boxes are used to enable the cables to be installed without damage (p. 53)

continuity of protective conductor test
a test to confirm that the circuit protective conductors are continuous throughout the circuit (p. 130)

continuity of ring final circuit conductors test
a test to confirm that the ring final circuit forms a continuous ring (p. 131)

correction factor
factors that have to be taken into account when selecting a cable for an installation (p. 31 on)

crampet
a simple type of "nail" type fixing for conduit (p. 56)

cross-sectional area of a conductor
the area of the cross-section when cut through a conductor (p. 16)

dado trunking
dado trunking is a form of trunking often used around open plan offices, generally mounted above desk height (p. 82)

discharge lamp
a lamp where the light is produced by current flowing through a gas or vapour (p. 105)

discrimination
discrimination should make sure that an overcurrent occurring at any point on the system causes the minimum disruption of supply (p. 117)

draw in point
a point in a conduit system where cables can be drawn in (p. 63)

earth electrode
a conductor that provides an electrical connection to earth (p. 120)

Economy 7 tariff
a tariff designed for electricity customers who use some electricity at night (p. 2)

elbow
a bend in conduit (p. 53)

electromagnetic effect
magnetic flux circulating in conduit or trunking (p. 46)

emergency lighting
can be on all the time (maintained) or only come on when the supply fails (non-maintained) (p. 102)

energy meter
or kilowatt hour meter, is a meter to measure the supply company's electrical energy supply to the customer (p. 1)

expansion couplings
a coupling used with PVC conduit to allow for expansion and contraction due to change in temperature (p. 62)

flexible conduit
conduit used where there may be movement or vibration (p. 67)

fire barrier
fire barriers are installed in trunking to prevent fire and smoke moving from one area to another (p. 79)

floor trunking
trunking laid on the floor so that the top edge is flush with the finished floor level (p. 84)

FP 200 cable
a fairly robust aluminium sheathed cable (pp. 12, 21)

fuse
a device to cut the circuit off from the supply when an overload occurs (p. 115)

galvanised conduit
conduit that has been cleaned and then dipped in hot zinc to provide a rust resistant finish (p. 50)

grommet strip
a plastic or neoprene strip which is pushed onto holes in trunking to prevent damage to the cable insulation (p. 79)

grouping
where several circuits are bunched together this can cause cables to overheat and go over their temperature limit, unless this factor is taken into account when selecting the size of cable (p. 34)

incandescent lamp
a lamp commonly used in the home which has a tungsten filament which when heated gives off a light (p. 104)

inspection box
a box in a conduit run that allows an inspection to be made of the cables installed (p. 53)

insulation
the insulation on a cable ensures that the current flows along its predetermined path and also prevents the possibility of electric shock (p. 7)

insulation resistance test
this test is to confirm that the insulation throughout the circuit has not been damaged (p. 131)

isolation
isolation means cutting off the installation, or circuit, from all sources of electrical supply to prevent danger (p. 113)

kilowatt hour
(kWh) the unit used for costing electrical energy (p. 1)

lighting trunking
trunking designed to support lighting fittings attached directly to the trunking (p. 84)

main equipotential bonding conductors
conductors installed to ensure that all exposed metalwork is at the same electrical potential (p. 121)

mechanical protection
protection for cables that can be by metal sheathing, conduit or trunking (p. 11)

MIMS
mineral-insulated metal-sheathed cable (p. 11)

mini-trunking
plastic trunking with a lid, of small physical size (p. 83)

oval conduit
conduit commonly used to protect wiring that is to be run flush into the building structure, restricted to use in straight lengths (p. 69)

overcurrent device
fuse or circuit breaker which operates when the circuit current exceeds the current rating (p. 33)

overload
a larger than normal current flowing in a circuit that is still electrically sound (p. 115)

oversheath
an outer covering on a cable (p. 12)

pipe vice
a pipe vice holds the conduit while it is cut to length (p. 51)

pliable and corrugated plastic conduit
this type of conduit has limitations, but a typical example for its use would be to connect a cooker outlet to a cooker panel in a stud partition wall (p. 68)

PME
protective multiple earthing, an earthing arrangement found in TN-C-S systems (p. 121)

polarity test
a polarity test checks that the installation connections have been made in the correct conductors (p. 132)

PVC
polyvinyl chloride – often used for cable insulation (p. 9)

radial circuit
is basically a loop in and loop out circuit (p. 107)

reamer
a deburring tool for conduit (p. 45)

reference method
a classification according to how a cable is to be installed (p. 32)

residual current device
RCD – a device which is able to detect very small earth fault currents and automatically disconnect the supply (p. 121)

resistivity
the ability of a material to resist the passage of an electric current (p. 9)

ring final circuit
where a circuit starts and finishes at the same point and forms a continuous ring (p. 107)

saddle
a type of fixing for conduit (p. 56)

setting block
a wooden setting block is used to bend conduit (p. 52)

sheathing
a protective covering on a cable (p. 9)

sherardized conduit

conduit that has been coated whilst hot with zinc dust which results in the steel becoming impregnated with zinc so that it does not flake off or crack (p. 50)

short circuit

when a fault occurs between live conductors (p. 116)

simmerstat

a temperature control device (p. 110)

space factor

the overall cross-sectional area of cables in conduit and trunking should not exceed 45% of the space available (p. 28)

spur

a branch cable run from a ring final circuit to one single or one twin socket outlet (p. 107)

standard rate tariff

a single rate tariff for electricity customers

stocks and dies

stocks and dies are used to cut a thread on a length of conduit (p. 50)

supports

are required for cables and enclosures (pp. 25, 57, 63, 79, 94, 98)

SWA

steel wire armoured cable widely used in industry (p. 21)

terminations

cable terminations should comply with certain conditions, for example, a termination should be mechanically and electrically sound (p. 28)

thermal insulation

where cables have to be totally enclosed in thermal insulation they must either be enclosed in conduit or trunking, or where this is unavoidable the appropriate factor must be applied. (pp. 33, 38)

thermostat

a temperature control device which measures the temperature and cuts off the supply when the required temperature is reached (p. 110)

trunking

trunking is used to provide good mechanical protection to conductors, often where a larger capacity is required than can be enclosed in conduit (p. 75)

tungsten halogen lamp

a lamp consisting of a tungsten filament in a thin glass tube which contains halogen gas (p. 105)